New Enhanced Sensitivity
Infrared Laser Spectroscopy Techniques
Applied to Reactive Plasmas and Trace Gas Detection

Inauguraldissertation

zur

Erlangung des akademischen Grades

doctor rerum naturalium (Dr. rer. nat.)

an der Mathematisch-Naturwissenschaftlichen Fakultät

der

Ernst-Moritz-Arndt-Universität Greifswald

vorgelegt von

Stefan Welzel

geboren am 6. Mai 1979

in Meerane

Greifswald, 9. Juni 2009

Bibliografische Information der Deutschen Nationalbibliothek

Die Deutsche Nationalbibliothek verzeichnet diese Publikation in der
Deutschen Nationalbibliografie; detaillierte bibliografische Daten sind
im Internet über http://dnb.d-nb.de abrufbar.

ISBN 978-3-8325-2345-9

Logos Verlag Berlin GmbH
Comeniushof, Gubener Str. 47,
10243 Berlin
Tel.: +49 (0)30 42 85 10 90
Fax: +49 (0)30 42 85 10 92
INTERNET: http://www.logos-verlag.de

Dekan: Prof. Dr. K. Fesser

1. Gutachter :   Prof. Dr. J. Röpcke

2. Gutachter: Prof. Dr. N. Sadeghi

Tag der Promotion: 9. Oktober 2009

Abstract

Infrared laser absorption spectroscopy (IRLAS) employing both tuneable diode and quantum cascade lasers (TDLs, QCLs) has been applied with both high sensitivity and high time resolution to plasma diagnostics and trace gas measurements.

TDLAS combined with a conventional White type multiple pass cell was used to detect up to 13 constituent molecular species in low pressure $Ar/H_2/N_2/O_2$ and $Ar/CH_4/N_2/O_2$ microwave discharges, among them the main products such as H_2O, NH_3, NO and CO, HCN respectively. The hydroxyl radical has been measured in the mid infrared (MIR) spectral range in-situ in both plasmas yielding number densities of between $10^{11} \dots 10^{12}$ cm^{-3}. Strong indications of surface dominated formation of either NH_3 or N_2O and NO were found in the $H_2 - N_2 - O_2$ system. In methane containing plasmas a transition between deposition and etching conditions and generally an incomplete oxidation of the precursor were observed.

The application of QCLs for IRLAS under low pressure conditions employing the most common tuning approaches has been investigated in detail. A new method of analysing absorption features quantitatively when the rapid passage effect is present is proposed. If power saturation is negligible, integrating the undisturbed half of the line profile yields accurate number densities without calibrating the system. By means of a time resolved analysis of individual chirped QCL pulses the main reasons for increased effective laser line widths could be identified. Apart from the well-known frequency down chirp non-linear absorption phenomena and bandwidth limitations of the detections system may significantly degrade the performance and accuracy of inter pulse spectrometers. The minimum analogue bandwidth of the entire system should normally not fall below 250 MHz.

QCLAS using pulsed lasers has been used for highly time resolved measurements in reactive plasmas for the first time enabling a time resolution down to about 100 ns to be achieved. A temperature increase of typically less than 50 K has been established for pulsed DC discharges containing Ar/N_2 and traces of NO. The main NO production and depletion reactions have been identified from a comparison of model calculations and time resolved measurements in plasma pulses of up to 100 ms. Considerable NO destruction is observed after $5 \dots 10$ ms due to the impact of N atoms.

Finally, thermoelectrically (TE) cooled pulsed and continuous wave (cw) QCLs have been employed for high finesse cavity absorption spectroscopy in the MIR. Cavity ring down spectroscopy (CRDS) has been performed with pulsed QCLs and was found to be limited by the intrinsic frequency chirp of the laser suppressing an efficient intensity build-up inside the cavity. Consequently the accuracy and advantage of an absolute internal absorption calibration is not achievable. A room temperature cw QCL was used in a complementary cavity enhanced absorption spectroscopy (CEAS) configuration which was equipped with different cavities of up to ~ 1.3 m length. This spectrometer yielded path lengths of up to 4 km and a noise equivalent absorption down to 4×10^{-8} $cm^{-1}Hz^{-1/2}$. The corresponding molecular concentration detection limit (e.g. for CH_4, N_2O and C_2H_2 at 1303 $cm^{-1}/7.66$ µm) was generally below 1×10^{10} cm^{-3} for 1 s integration times and one order of magnitude less for 30 s integration times. The main limiting factor for achieving even higher sensitivity is the residual mode noise of the cavity. Employing a ~ 0.5 m long cavity the achieved sensitivity was good enough for the selective measurement of trace atmospheric constituents at 2.2 mbar.

Kurzfassung

Im Rahmen dieser Arbeit wurden hochempfindliche und hochzeitaufgelöste Untersuchungen an reaktiven molekularen Plasmen und Spurengasmessungen mittels Infrarotlaser-Absorptionsspektroskopie (IRLAS) durchgeführt. Als Lichtquellen fanden durchstimmbare Diodenlaser (Bleisalzlaser) und die erst seit einigen Jahren verfügbaren Quantenkaskadenlaser (engl. Abk.: QCL) Verwendung.

IRLAS mittels Bleisalzlasern wurde angewendet, um molekulare Teilchendichten von bis zu 13 verschiedenen Molekülen in $Ar/H_2/N_2/O_2$- und $Ar/CH_4/N_2/O_2$-Mikrowellenplasmen zu quantifizieren. Hierzu kam eine optische Langwegzelle (sog. White-Zelle) mit 60 m Absorptionsweg innerhalb des Niederdruckreaktors zum Einsatz. Das OH-Radikal konnte erstmals mit dieser Technik im mittleren Infrarotbereich (MIR, 3 - 20 µm) mit Teilchendichten von 10^{11} ... 10^{12} cm^{-3} nachgewiesen werden. Weiterhin fanden sich deutliche Hinweise auf eine von Oberflächenreaktionen dominierte Bildung der Hauptprodukte NH_3 bzw. N_2O und NO im H_2 - N_2 - O_2-Modellsystem. Für methanhaltige Plasmen wurde ein Übergang von beschichtenden zu schichtabtragenden Bedingungen sowie eine generell unvollständige Oxidation von CH_4 (d.h. ein überwiegender Anteil von CO statt CO_2 in der Gasphase) festgestellt.

Die Anwendung bekannter Durchstimmverfahren für QCLs wurde hinsichtlich ihrer Anwendbarkeit unter Niederdruckbedingungen detailliert untersucht. Dabei konnte ein Ansatz zur quantitativ korrekten und kalibrierfreien Auswertung verfälschter Absorptionslinien entwickelt werden. Die automatische schnelle Frequenzdurchstimmung in gepulsten QCLs führt zu nichtlinearen Absorptionserscheinungen (d.h. Sättigung und schwach gedämpfte Populationsschwankungen der betroffenen Übergänge). Sofern keine Sättigungserscheinungen vorliegen, lässt sich aus dem ungestörten Teil des Linienprofils (kalibrierfrei) eine Teilchendichte ableiten, deren systematischer Fehler unterhalb der üblichen Messfehler liegt. Weiterhin wurden einzelne Laserpulse während der Durchstimmung des QCLs mittels Stromrampe zeitaufgelöst analysiert. Neben der automatischen Frequenzdurchstimmung werden Genauigkeit und Empfindlichkeit der Methode stark durch die genannten nichtlinearen Absorptionseffekte und üblicherweise vernachlässigte Bandbreitenbeschränkungen bei Detektor, Vorverstärker oder Datenerfassungskarte reduziert. Im Spektrum äußert sich dies durch asymmetrische Linien und eine scheinbar extrem große effektive Laserlinienbreite. Basierend auf der Durchstimmrate des Lasers wird eine Abschätzung für die minimal erforderliche analoge Bandbreite des Gesamtsystems angegeben, die typischerweise 250 MHz nicht unterschreiten sollte.

Basierend auf den Voruntersuchungen zu gepulsten QCLs konnten zeitaufgelöste IRLAS Messungen in einer gepulsten DC-Entladung durchgeführt werden. Die Zeitauflösung war nach unten durch die Laserpulsweite von ca. 100 ns beschränkt. Für gepulste Ar/N_2 Entladungen mit maximal 1 % NO-Zumischung wurde ein Anstieg der Gastemperatur von maximal 50 K beobachtet. Auf Grundlage einer zeitabhängigen Modellrechnung konnten Hauptreaktionspfade bezüglich NO in bis zu 100 ms langen Plasmapulsen ermittelt werden. Ein merklicher Abbau von NO wird nach 5 ... 10 ms beobachtet und hauptsächlich durch Gasphasenreaktionen mit N-Atomen hervorgerufen.

Im letzten Teil der Arbeit wurde das Potential zur Kombination leistungsstarker, thermoelektrisch gekühlter QCLs mit optischen Resonatoren untersucht. Hierbei kamen gepulste Laser (für die Cavity-Ring-Down-Spektroskopie - CRDS) und kontinuierlich betriebene (cw) QCLs (für die Cavity-Enhanced-Absorptions-Spektroskopie) im MIR zum Einsatz. Die Frequenzdurchstimmung gepulster QCLs verhindert ebenfalls einen sinnvollen Einsatz im Rahmen der (gepulsten) CRDS, da die typische Intensitätsverstärkung auf den Resonatormoden ausbleibt. Zusammen mit ungünstigen Bandbreiteneffekten verliert die CRDS den Vorteil einer kalibrierfreien, empfindlichen Messmethode. Im Gegensatz dazu konnten cw QCLs (betrieben bei Raumtemperatur) erfolgreich mit der CEAS-Methode eingesetzt werden. Mit Resonatorlängen bis zu 1.3 m wurden effektive Absorptionsweglängen von bis zu 4 km und eine bandbreitennormierte Empfindlichkeit von $4 \times 10^{-8} \, cm^{-1} Hz^{-1/2}$ erreicht werden. Die sich daraus ergebenden Detektionsgrenzen für diverse Moleküle (z.B. CH_4, N_2O oder C_2H_2 bei 1303 cm^{-1}/7.66 µm) wären geringer als $1 \times 10^{10} \, cm^{-3}$ bei einer Mittelungszeit von 1 s. Im Falle einer Mittelungszeit von 30 s würde die Detektionsgrenze um eine weitere Größenordnung sinken. Die Hauptbegrenzung der Empfindlichkeit beruht auf dem sog. Modenrauschen, welches typisch für CEAS-Anwendungen ist. Trotzdem genügte ein nur ca. 0.5 m langer Resonator (etwa 0.3 l Volumen), um die atmosphärischen Spurengasbestandteile bei reduziertem Druck (2.2 mbar) bequem nachweisen zu können.

Content

Abstract 5

Kurzfassung 7

Content 9

1 Introduction 13
  1.1 Reactive Plasmas 13
  1.2 Plasma diagnostics 14
  1.3 Challenges in infrared laser absorption spectroscopy 15
  1.4 Trace gas detection 17
  1.5 Outline of the thesis 18
  Bibliography 18

2 Infrared absorption spectroscopy 23
  2.1 Mid infrared light sources 23
    2.1.1 Overview 23
    2.1.2 Tuneable diode lasers 25
    2.1.3 Quantum cascade lasers 27
  2.2 Optical absorption theory 30
    2.2.1 Linear absorption 30
    2.2.2 Line profiles and spectral broadening 31
    2.2.3 Infrared absorption cross sections and databases 34
    2.2.4 Line strength and molecular line parameters 35
    2.2.5 Non-linear absorption 37
  2.3 Sensitivity considerations 40
  A Appendix 43
    A.1 Spectral coverage of selected infrared light sources 43
    A.2 Line profiles 44
    A.3 Line strength and transition probabilities 45
  Bibliography 47

3 Molecule conversion in reactive plasmas containing H_2-N_2-O_2 53
  3.1 Motivation 53
  3.2 Experimental 55
    3.2.1 Discharge setup 55
    3.2.2 Injected precursors 56
    3.2.3 Spectroscopic issues 57
  3.3 Results and discussion 58
    3.3.1 Discharge parameters 58
    3.3.2 (Ar -) H_2 - N_2 - O_2 discharges 59
    3.3.3 (Ar -) CH_4 - N_2 - O_2 discharges 62
  3.4 Summary and conclusions 65
  Bibliography 67

4   Quantum cascade laser absorption spectroscopy   **71**

4.1   Introduction and motivation   **71**

4.2   Comparison between QCLAS and TDLAS   **73**

4.2.1   Modes of operation   73

4.2.2   Experimental arrangement   75

4.2.3 CH_4 detection at elevated pressure   78

4.2.4   CO detection at low pressure   80

4.3   Properties of chirped QCL pulses   **85**

4.3.1   Chirp rate of pulsed QCLs   85

4.3.2   Adiabatic and linear rapid passage   87

4.3.3   Resolution and bandwidth limits employing pulsed QCLs   89

4.4   Bandwidth effects using short QCL pulses   **91**

4.4.1   Overview of measurement systems in the literature   91

4.4.2   Experimental approach   93

4.4.3   Concentration measurements using the conventional *inter* pulse
        technique   94

4.4.4   Consequences of the frequency chirp   96

4.4.5   Bandwidth effects   100

4.5   Discussion and conclusions   **104**

B   Appendix   **110**

Bibliography   **112**

**5   Time resolved study of a pulsed DC discharge using
       quantum cascade laser absorption spectroscopy**   **117**

5.1   Introduction   **117**

5.2   Experimental   **119**

5.2.1   Discharge setup   119

5.2.2   Spectroscopic setup   120

5.3   Characterisation of the arrangement   **122**

5.3.1   Electric parameters   122

5.3.2   Spectroscopic issues   123

5.4   Results and discussion   **125**

5.4.1   NO depletion in a single 1 ms plasma pulse   125

5.4.2   NO depletion with multiple plasma pulses   126

5.4.3   NO depletion in a single 100 ms plasma pulse   127

5.4.4   Temperature evolution during a plasma pulse   128

5.4.5   Influence of the temperature on the spectroscopic results   130

5.4.6   Heavy species kinetics in pulsed DC discharges containing N_2 132

5.5   Summary and conclusions   **134**

C   Appendix   **135**

C.1   Temperature evolution in a DC discharge tube   **135**

C.2   Supplemental figures for long plasma pulses   **136**

Bibliography   **137**

6   Trace gas measurements using optically resonant cavities and quantum cascade lasers   **139**

6.1   Introduction   139

6.2   Theoretical considerations   141
    6.2.1   Cavity ring down effect   142
    6.2.2   Cavity enhanced absorption   143
    6.2.3   Integrated cavity output   144
    6.2.4   Detection limits   144

6.3   Experimental   145
    6.3.1   Pulsed cavity excitation   145
    6.3.2   Continuous cavity excitation   147

6.4   Results and Discussion   149
    6.4.1   CRDS and ICOS using a pulsed QCL   149
        6.4.1.1 CH_4 detection at 7.42 μm   149
        6.4.1.2 N_2O detection at 8.35 μm   150
        6.4.1.3   Sensitivity conclusions   152
    6.4.2   Bandwidth effects with a pulsed QCL and optical cavities   153
        6.4.2.1   Determination of the frequency chirp   153
        6.4.2.2   Consequences of the chirped pulse for optical cavities   154
    6.4.3   CEAS using a cw QCL   156
        6.4.3.1   Calibration of the method and test of validity   156
        6.4.3.2   System performance   160
        6.4.3.3   Sensitivity improvements   161
        6.4.3.4   Discussion of sensitivity achievements   164

6.5   Conclusions   165

D   Appendix   **167**

Bibliography   **169**

7   Accuracy and limitations   **173**

8   Summary and outlook   **179**

8.1   Summary   179

8.2   Outlook   181

9   Glossary   **183**

9.1   Acronyms   183

9.2   Symbols   184

Acknowledgement   **187**

Curriculum Vitae   **189**

Publications   **191**

Declaration   **197**

1   Introduction

1.1   Reactive Plasmas

Non-thermal molecular plasmas are of increasing interest not only as model systems in fundamental research but also in plasma processing and technology. They are characterised by a non-equilibrium between a relatively low neutral gas temperature in contrast to a significantly higher kinetic temperature of the electrons. Such reactive plasmas are used in a variety of applications among them thin film deposition and etching, compound and molecule synthesis, surface modification and functionalisation, semiconductor processing and medical applications, such as decontamination [1,2]. Recent concerns about air pollution and greenhouse gases have led to the application of plasma assisted waste treatment and volatile organic compound removal strategies [3-6]. Furthermore reactive plasmas are present in fusion reactors [7,8] or during the re-entrant of space vehicles [9].

This broad range of applications motivate fundamental investigations of these complex plasmas that are usually accomplished by combining plasma diagnostics and theoretical modelling. Apart from the discharge physics the plasma chemistry and reaction kinetics is of special interest in reactive plasmas. Figure 1.1 provides an overview of typical time scales in reactive non-equilibrium discharges covering about 10 orders of magnitude from their ignition to long-time plasma surface interactions. However, even for only a limited number of molecular precursors (e.g., N_2, O_2, H_2 or binary mixtures of them) numerous processes have to be accounted for and require a thorough theoretical approach [9,12-15]. More complex gas mixtures that are used in practise, e.g., those containing hydrocarbons, fluorocarbons, silane, organo-silicon or boron compounds, increase the number of involved reactions and thus the complexity of the models inevitably [16-20]. Alternatively, an approach based on chemical quasi-equilibria and macroscopic net reactions has been proposed and verified for different plasma chemical systems [21].

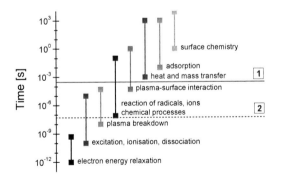

Figure 1.1:
Typical time scales of relevant processes in reactive non-thermal plasmas. (adapted from [10,11]). Present (1) and envisaged (2) time resolution limits of infrared laser based techniques are indicated for comparison.

The majority of detailed (microscopic) models concentrate on gas phase reactions for the description of the plasma chemistry. Specifically under low pressure conditions the probability of molecule formation by three-body association is small [22] and surface association mechanisms have to be considered. The non-negligible influence of the surface material for the ammonia production in discharges has been known for a long time [23].

However, the importance of surface production of molecules and their implementation into modelling has only recently been recognised [15,22,24-30]. Consequently, plasma catalytic approaches have increasingly been used [6,10,31-33]. Non-thermal plasmas at atmospheric pressures or discharges in liquids in general, in or in contact with air in particular have, attracted considerable interest [34,35]. Although especially N_2/O_2 mixtures have extensively been investigated as model systems [9,25], the complex plasma chemical behaviour in practise, specifically in the presence of water vapour, is still not well understood [36,37] and requires both additional theoretical and experimental studies. However, the investigation of atmospheric pressure discharges is challenging because of transient filaments with strong gradients and a lifetime of few tens of nanoseconds. A more feasible approach in respect to plasma surface interactions is therefore to study low pressure discharges which may be relatively homogeneous in large volumes [38].

A key to improved understanding of fundamental phenomena in molecular non-thermal plasmas is the measurement of transient or stable plasma reaction products, i.e. detecting and monitoring their ground state concentrations, which can be used in turn for plasma chemical modelling. Due to longer lifetimes of stable molecules their density is typically higher compared to radicals. Information about processes can thus indirectly be obtained by measuring the easier accessible stable products. Schram recently pointed out that not only densities, but also lifetimes, i.e. their ratio namely the production and loss of radicals, are relevant [22]. Hence, a high (and easier to measure) density of a specific radical may not be an indicator of its high reactivity or importance. Consequently, the sensitivity requirements for diagnostic techniques for detecting highly reactive species are even stronger (section 1.3).

Since surface related processes typically occur on longer time scales (figure 1.1), pulsed discharges can facilitate a discrimination between surface and gas phase reactions [38,39]. Pulsing is also often employed as an additional opportunity for process control, since plasma parameters, e.g. densities and temperatures, become time dependent [40,41]. Moreover, a self-pulsing regime was observed in microplasmas [42,43]. Time dependent plasma properties therefore impose an additional selection criterion for their sensing, namely the time resolution of the individual diagnostic technique.

1.2 Plasma diagnostics

The need for a better scientific understanding of plasma physics and chemistry has stimulated the improvement of established diagnostic techniques and the introduction of new ones. An ideal technique would have the following main attributes:

i) non-invasive to obviate perturbations of the discharge,
ii) *in-situ* approach without the need for extracting samples,
iii) local measurement technique for achieving good spatial resolution,
iv) selective and only sensitive to the target parameter or species,
v) highly sensitive to small quantities,
vi) a fast response time to provide time-resolved measurements,
vii) covering several orders of magnitude in both time and concentration (i.e. high dynamic range, figures 1.1 and 1.2), and
viii) yielding absolute quantities without additional calibration.

It is clear that these desirable properties are not available simultaneously in a single device. Therefore different diagnostic techniques have been applied to reactive plasmas [44-48], of which only a few (i.e., the most widespread ones) are mentioned below.

Electrical probes are employed to obtain electron densities or information on the electron energy distribution function (EEDF) and the plasma potential, respectively. This method is clearly invasive and especially difficult to implement in harsh or depositing chemical environments. Gas chromatography (GC) and mass spectrometry (MS) are often used for identification and sensitive quantification of plasma species. However, both are based on an *ex-situ* sampling approach and require calibration. Additionally, the response time of GC is usually in the order of minutes and the selectivity of MS in complex gas mixtures may be limited. Methods based on traditional optical spectroscopy have now become amongst the most important, since they provide non-invasive, *in-situ* diagnostics with a sufficient time resolution. Optical emission spectroscopy (OES) yields line of sight measurements on electronically excited states of species. Hence, collision-radiation models, knowledge of the EEDF and a complete instrument calibration are necessary for retrieving ground state concentrations. Laser induced fluorescence (LIF) yields localised information on ground state number densities, but is also not calibration free. Finally, optical absorption spectroscopy (OAS) provides a means of determining the population densities of species in both ground and excited states. Species identification is deduced from the spectral line positions while line profiles are often connected with gas temperature and relative intensities provide information about population densities. An important advantage of OAS over OES methods is that only relative intensities need to be measured. Continuous light sources or broadband frequency combs in combination with dispersive elements or a Michelson interferometer (Fourier transform spectroscopy, FTS) as well as tuneable narrow-band light sources can be employed [45-50]. OAS has been applied right across the spectrum from the vacuum ultraviolet (VUV) to the far infrared. The UV/VUV spectral range is of special interest for the detection of atomic species and molecular radicals in plasmas, although the selectivity for radicals may be limited due to overlapping absorption features and the achievable resolution of spectrometers. The increasing availability of compact, adequately powerful, tuneable and narrow line width light sources along with appropriate detectors has now made the near and mid infrared (NIR, MIR) spectral range attractive for measuring molecular species by means of their overtone or fundamental transitions.

1.3   Challenges in infrared laser absorption spectroscopy

Infrared laser absorption spectroscopy (IRLAS) is a non-invasive and selective means for measuring *in-situ* line of sight averaged concentrations of free radicals, transient molecules and stable products in their electronic ground states. Conversely, it is also applied in purely spectroscopic studies to extract structural information and basic spectroscopic data of molecular species, such as molecular line parameters, transition dipole moments and line strengths, respectively [51-53].

Such high resolution absorption cross section or line strength data, usually being collected in spectroscopic databases, are essential for obtaining absolute number densities from relative intensity measurements. Additionally, the temperature dependence of absorption cross sections have to be known and taken into account which is especially important for plasma diagnostics [40] where the gas temperature is typically different from the reference temperature (296 K) of the database[1]. In some cases the knowledge of the relative change of the line strength with the temperature enables the gas temperature to be abstracted from only a single absorption line [40] or a line ratio [54] rather than from a conventional Boltzmann plot encompassing considerably more lines.

IRLAS has now been established as a versatile tool for process monitoring and developing a deeper understanding of a variety of plasma chemical systems containing hydrocarbons [16,17,38,54-61], fluorocarbons [24,39,62-64], nitrogen, oxygen and/or hydrogen mixtures [29,65,66], silane [56,67] or organo-silicon [68] compounds. Other examples and an extensive list of references are given in [47,69].

The majority of studies focussed on transient molecules and stable products which can already provide valuable information on the chemical activity of the plasmas [22]. Typical absorption cross sections in the MIR enable the detection of number densities down to 10^{13} ... 10^{14} cm^{-3} in single pass configuration corresponding to concentrations in the per cent and per mille range for low pressure plasmas (figure 1.2). In other words, the depletion of the precursors and production of main stable species are readily accessible. Minor products, intermediate molecules and radicals of sufficient lifetime can be detected by means of increased absorption paths down to 10^{10} ... 10^{11} cm^{-3} (figure 1.2). However, important radicals are usually predicted at even lower concentrations [16,61,70-72]. Apart from high resolution spectroscopic data which is often absent for short-lived species, the achievable sensitivity is a key challenge for in-situ IRLAS and thus highly sensitive extensions of OAS are desirable. Common approaches of achieving a better signal-to-noise ratio to increase the sensitivity is the reduction of the noise level by means of modulation techniques or averaging. Alternatively, a higher fractional absorption is accomplished by multiple pass cells or optical resonators. Particularly, optical cavities provide a better spatial resolution, i.e. more localised measurements, than long path cell arrangements of the same absorption length.

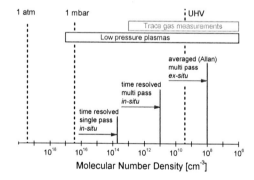

Figure 1.2:
Overview of expected number densities in reactive low pressure plasmas and for trace gas detection. Typical limits of common IRLAS detection are marked by arrows. The number densities corresponding to atmospheric pressure, 1 mbar und the ultra high vacuum (UHV) range are indicated (dotted).

Studying the reaction kinetics of molecular plasmas generally requires - apart form surface related processes - a time resolution below 1 s. Tuneable diode laser (TDL) systems can achieve ms time resolution (marker (1) in figure 1.1) [73], without averaging even sub-ms time resolution is possible. However, not all chemical processes are entirely covered by such

---

[1] See chapter 2 (2.2.3) for detailed information and references concerning spectral databases.

a data acquisition speed. The application of pulsed quantum cascade lasers (QCLs) provide full spectral scans of up to 1 cm^{-1} in about 100 ns. Hence, a time resolution of the same order has become possible for quantitative *in-situ* measurements of molecular concentrations for the first time. Therefore it fits very well to measurements of rapidly changing chemical processes in plasmas (marker (2) in figure 1.1).

Unfortunately the main challenges in IRLAS, i.e. achieving high sensitivity and time resolution at the same time, are somewhat diametrically opposed. So far only *ex-situ* sampling in astigmatic Herriott cells of absorption paths exceeding 100 m and extensive averaging up to minutes to reach the Allan variance minimum has facilitated molecular detection limits below 10^{10} cm^{-3}. Such sensitivities are clearly superior to time resolved *in-situ* approaches (figure 1.2). This may be overcome by employing optical resonators yielding similar or longer *in-situ* absorption paths. On the other hand, this affects the achievable time resolution: the intensity of a single cavity mode decays within several ten µs. In other words, the acquisition time of a reasonably averaged spectrum consisting of a scan across about 250 of such cavity modes approaches the s range. Even at a constant spectral position the time resolution can never fall below the cavity decay time (~ 10 µs) which in turn is much longer than predicted for the application of pulsed QCLs for *in-situ* diagnostics. Adjusting either the detection limit or the time resolution will therefore always degrade the system performance in respect to the second criterion.

1.4   Trace gas detection

Although the MIR spectral range is well-known as the molecular fingerprint region, the application of IRLAS has so far been limited to niche applications in research laboratories. Apart from plasma diagnostics the most prominent research fields were atmospheric trace gas sensing, specifically isotope ratio measurements [74], breath gas analysis [75] or astrophysics [53]. The advent of QCLs and their commercial availability have led to increasing interest in MIR chemical sensing. In contrast to TDLs (i.e., lead salt lasers) there is no need for cryogenic cooling. This has supported the trend to field applications in conventional trace gas sensing [76,77] and to the development of new laser based sensors for explosive detection or as watchdogs for harmful chemical substances [78-80]. Common to all applications are only small temperature changes compared to plasmas and the need for moderate time resolutions which release some of the above mentioned challenges. Trace gas detection focuses mainly on highly sensitive and precise approaches. The combination of optical resonators with compact MIR lasers thus seems to be ideally suited for decreasing the detection limit further (figure 1.2) while reducing the necessary sampling volume and shrinking the apparatus dimensions in potentially field deployable systems at the same time.

1.5   Outline of the thesis

The main objective of this work is the development of advanced MIR absorption techniques by exploiting novel types of tuneable semiconductor based lasers, namely QCLs, and their application for both highly time resolved and highly sensitive measurements, particularly focussing on reactive plasmas.

The thesis is virtually organised in three parts, a theoretical (chapter 2) and two experimental (chapters 3 and 4 - 6) ones. In chapter 2 the current status of semiconductor based MIR laser light sources as well as theoretical and sensitivity considerations concerning IRLAS are presented.

Part one of the experimental work (chapter 3) concerns specific aspects of molecule conversion in $Ar/CH_4/N_2/O_2$ and $Ar/H_2/N_2/O_2$ microwave plasmas being, in fact, an extension of earlier investigations [81,82]. However, the focus of the measurements by means of lead salt lasers was now on the detection of the OH radical and finding evidence of plasma-surface interactions providing thereby a link to studies in an expanding thermal plasma [28,29].

The second experimental part concentrates on advanced QCL instrumentation. After an assessment of the accuracy of available QCL tuning mechanisms (chapter 4), chapter 5 follows with time resolved measurements of the decay of small admixtures of NO in a pulsed DC discharge. The temperature evolution and main chemical reactions in the discharge are in the centre of interest. In chapter 6 the combination of QCLs with optically resonant cavities is investigated and trace gas measurements by means of cavity enhanced absorption spectroscopy are presented. Finally, the accuracy and limitations of the applied methods are critically discussed (chapter 7). Conclusions and an outlook are presented in chapter 8.

Bibliography

[1]   K. Muraoka, M. Maeda, *Laser-Aided Diagnostics of Plasmas and Gases*, ISBN 0-7503-0643-2, (IOP Publishing, London, 1999).

[2]   H. Yasuda, *Plasma Polymerization*, ISBN 0-12-768760-2, (Academic Press, London, 1985).

[3]   *Non-Thermal Plasma Techniques for Pollution control*, eds. B. M. Penetrante, S. E. Schultheis, ISBN 0-387-57174-4, (Springer, New York, 1993).

[4]   J.S. Chang, *Science and Technology of Advanced Materials* **2**, 571 (2001).

[5]   J.S. Chang, *Plasma Sources Sci. Technol.* **17**, 045004 (2008).

[6]   U. Roland, F. Holzer, F.D. Kopinke, *Cat. Today* **73**, 315 (2002).

[7]   F.L. Tabares, R. Rohde, ASDEX Upgrade Team, *Plasma Phys. Control. Fusion* **46**, B381 (2004).

[8]   G.F. Counsell, *Plasma Sources Sci. Technol.* **11**, A80 (2002).

[9]   M. Capitelli, C.M. Ferreira, B.F. Gordiets, A.I. Osipov, *Plasma Kinetics in Atmospheric Gases*, ISBN 3-540-67416-0, (Springer, Berlin, 2000).

[10]   T. Nozaki, N. Muto, S. Kado, K. Okazaki, *Cat. Today* **89**, 57 (2004).

[11]   H.E. Wagner, R. Brandenburg, K.V. Kozlov, A. Sonnenfeld, P. Michel, J.F. Behnke, *Vacuum* **71**, 417 (2003).

[12]   K. Dittmann, D. Drozdov, B. Krames, J. Meichsner, *J. Phys. D: Appl. Phys.* **40**, 6593 (2007).

[13]   C.D. Pintassilgo, O. Guaitella, A. Rousseau, *Plasma Sources Sci. Technol.* **18**, 025005 (2009).

[14]   B. Gordiets, C.M. Ferreira, M.J. Pinheiro, A. Ricard, *Plasma Sources Sci. Technol.* **7**, 363 (1998).

[15]   B. Gordiets, C.M. Ferreira, M.J. Pinheiro, A. Ricard, *Plasma Sources Sci. Technol.* **7**, 379 (1998).

[16]   F. Hempel, P.B. Davies, D. Loffhagen, L. Mechold, J. Röpcke, *Plasma Sources Sci. Technol.* **12**, S98 (2003).

[17]   L. Mechold, J. Röpcke, X. Duten, A. Rousseau, *Plasma Sources Sci. Technol.* **10**, 52 (2001).

[18]   W.Y. Fan, P.F. Knewstubb, M. Käning, L. Mechold, J. Röpcke, P.B. Davies, *J. Phys. Chem. A* **103**, 4118 (1999).

[19]   K. Hassouni, O. Leroy, S. Farhat, A. Gicquel, *Plasma Chem. Plasma Process.* **18**, 325 (1998).

[20]   M.J. Kushner, *J. Appl. Phys.* **63**, 2532 (1988).

[21]   A. Rutscher, H.E. Wagner, *Plasma Sources Sci. Technol.* **2**, 278 (1993).

[22]   D.C. Schram, *Plasma Sources Sci. Technol.* **18**, 014003 (2009).

[23]   E.N. Eremin, A.N. Maltsev, V.L. Syaduk, *Russ. Journ. Phys. Chem.* **45**, 635 (1971).

[24]   J.A. O'Neill, Jyothi Singh, *J. Appl. Phys.* **77**, 497 (1995).

[25]   B. Gordiets, C.M. Ferreira, J. Nahorny, D. Pagnon, M. Touzeau, M. Vialle, *J. Phys. D: Appl. Phys.* **29**, 1021 (1996).

[26]   G. Kokkoris, A. Goodyear, M. Cooke, E. Gogolides, *J. Phys. D: Appl. Phys.* **41**, 195211 (2008).

[27]   G. Kokkoris, A. Panagiotopoulos, A. Goodyear, M. Cooke, E. Gogolides, *J. Phys. D: Appl. Phys.* **42**, 055209 (2009).

[28]   J.H. van Helden, Ph.D. Thesis, Eindhoven University of Technology, ISBN 978-90-386-2511-9, 2006.

[29]   R.A.B. Zijlmans, Ph.D. Thesis, Eindhoven University of Technology, ISBN 978-90-386-1288-1, 2008.

[30]   D.C. Schram, R.A.B. Zijlmans, J.H. van Helden, O. Gabriel, G. Yagci, S. Welzel, J. Röpcke, R. Engeln, *Journ. Optoelectronics Adv. Mater.* **10**, 1904 (2008).

[31]   P. Vankan, T. Rutten, S. Mazouffre, D.C. Schram, R. Engeln, *Appl. Phys. Lett.* **81**, 418 (2002).

[32]   A.M. Harling, D.J. Glover, J.C. Whitehead, K. Zhang, *Environ. Sci. Technol.* **42**, 4546 (2008).

[33]   F. Thevenet, O. Guaitella, E. Puzenat, J.M. Herrmann, A. Rousseau, C. Guillard, *Cat. Today* **122**, 186 (2007).

[34]   U. Kogelschatz, B. Eliasson, W. Egli, *Pure Appl. Chem.* **71**, 1819 (1999).

[35]   P. Bruggeman, C. Leys, *J. Phys. D: Appl. Phys.* **42**, 053001 (2009).

[36]   F. Thevenet, O. Guaitella, E. Puzenat, C. Guillard, A. Rousseau, *Appl Cat. B: Environm.* **84**, 813 (2008).

[37]   A.V. Pipa, J. Röpcke, *IEEE Trans. Plasma Sci.*, ISSN 0093-3813, accepted (2009).

[38]   A. Rousseau, O. Guaitella, L. Gatilova, M. Hannemann, J. Röpcke, *J. Phys. D: Appl. Phys.* **40**, 2018 (2007).

[39]   O. Gabriel, S. Stepanov, J. Meichsner, *J. Phys. D: Appl. Phys.* **40**, 7383 (2007).

[40]   O. Gabriel, S. Stepanov, M. Pfafferott, J. Meichsner, *Plasma Sources Sci. Technol.*
        15, 858 (2006).
[41]   A. Rousseau, E. Teboul, N. Sadeghi, *Plasma Sources Sci. Technol.* **13**, 166 (2004).
[42]   X. Aubert, G. Bauville, J. Guillon, B. Lacour, V. Puech, A. Rousseau, *Plasma
        Sources Sci. Technol.* **16**, 23 (2007).
[43]   X. Aubert, A. Pipa, J. Röpcke, A. Rousseau, presented at the 28th ICPIG, Prague,
        Czech Republic, 15-20 July 2007, ISBN 978-80-87026-01-4 (3P10).
[44]   J.M. Stillahn, K.J. Trevino, E.R. Fisher, *Annu. Rev. Anal. Chem.* **1**, 261 (2008).
[45]   N. Hershkowitz, R.A. Breun, *Rev. Sci. Instrum.* **68**, 880 (1997).
[46]   R.A. Gottscho, T.A. Miller, *Pure Appl. Chem.* **56**, 189 (1984).
[47]   M. Lackner, *Rev. Chem. Eng.* **23**, 65 (2007).
[48]   M. Hori, T. Goto, *Plasma Sources Sci. Technol.* **15**, S74 (2006).
[49]   K. Tachibana, *Plasma Sources Sci. Technol.* **11**, A166 (2002).
[50]   M.J. Thorpe, F. Adler, K.C. Cossel, M.H.G. de Miranda, J. Ye, *Chem. Phys. Lett.*
        468, 1 (2009).
[51]   P.B. Davies, *Spectrochim. Acta Part A* **55**, 1987 (1999).
[52]   G.D. Stancu, J. Röpcke, P.B. Davies, *Journ. Chem. Phys.* **122**, 014306 (2005).
[53]   G. Winnewisser, T. Drascher, T. Giesen, I. Pak, F. Schmülling, R. Schieder,
        Spectrochim. Acta Part A **55**, 2121 (1999).
[54]   J. Hirmke, F. Hempel, G.D. Stancu, J. Röpcke, S.M. Rosiwal, R.F. Singer, *Vacuum*
        80, 967 (2006).
[55]   J. Hirmke, A. Glaser, F. Hempel, G.D. Stancu, J. Röpcke, S.M. Rosiwal, R.F. Singer,
        Vacuum **81**, 619 (2007).
[56]   P.B. Davies, P.M. Martineau, *Adv. Mater.* **4**, 729 (1992).
[57]   J. Röpcke, L. Mechold, M. Käning, W.Y. Fan, P.B. Davies, *Plasma Chem. Plasma
        Process.* **19**, 395 (1999).
[58]   J. Röpcke, L. Mechold, X. Duten, A. Rousseau, *J. Phys. D: Appl. Phys.* **34**, 2336
        (2001).
[59]   G. Lombardi, K. Hassouni, G.D. Stancu, L. Mechold, J. Röpcke, A. Gicquel, *Plasma
        Sources Sci. Technol.* **14**, 440 (2005).
[60]   C. Busch, I. Möller, H. Soltwisch, *Plasma Sources Sci. Technol.* **10**, 250 (2001).
[61]   I. Möller, A. Serdyuchenko, H. Soltwisch, *J. Appl. Phys.* **100**, 033302 (2006).
[62]   M. Haverlag, E. Stoffels, W.W. Stoffels, G.M.W. Kroesen, F.J. de Hoog, *J. Vac. Sci.
        Technol. A* **12**, 3102 (1994)
[63]   D.B. Oh, A.C. Stanton, H.M. Anderson, M.P. Splichal, *J. Vac. Sci. Technol. B* **13**,
        954 (1995).
[64]   K. Miyata, H. Arai, M. Hori, T. Goto, *J. Appl. Phys.* **82**, 4777 (1997).
[65]   L.V. Gatilova, K. Allegraud, J. Guillon, Y.Z Ionikh, G. Cartry, J. Röpcke, A.
        Rousseau, *Plasma Sources Sci. Technol.* **16**, S107 (2007).
[66]   Y. Ionikh, A.V. Meshchanov, J. Röpcke, A. Rousseau, *Chem. Phys.* **322**, 411 (2006).
[67]   T. Goto, *Pure Appl. Chem.* **68**, 1059 (1996).
[68]   J. Röpcke, G. Revalde, M. Osiac, K. Li, J. Meichsner, *Plasma Chem. Plasma
        Process.* **22**, 137 (2002).
[69]   J. Röpcke, G. Lombardi, A. Rousseau, P.B. Davies, *Plasma Sources Sci. Technol.*
        15, S148 (2006).
[70]   K. Matyash, R. Schneider, A. Bergmann, W. Jacob, U. Fantz, P. Pecher, *Journ. Nucl.
        Mater.* **313-316**, 434 (2003).

[71] K. Hassouni, X. Duten, A. Rousseau, A. Gicquel, *Plasma Sources Sci. Technol.* **10**, 61 (2001).

[72] G. Lombardi, K. Hassouni, G.D. Stancu, L. Mechold, J. Röpcke, A. Gicquel, *J. Appl. Phys.* **98**, 053303 (2005).

[73] J.B. McManus, D. Nelson, M. Zahniser, L. Mechold, M. Osiac, J. Röpcke, A. Rousseau, *Rev. Sci. Instrum.* **74**, 2709 (2003).

[74] E. Kerstel, L. Gianfrani, *Appl. Phys. B* **92**, 439 (2008).

[75] M. Mürtz, P. Hering, Physik Journ. **7** (10), 43 (2008). (*in German*)

[76] M.R. McCurdy, Y. Bakhirkin, G. Wysocki, R. Lewicki, F.K. Tittel, *J. Breath Res.* **1**, 014001 (2007).

[77] B. Tuzson, J. Mohn, M.J. Zeeman, R.A. Werner, W. Eugster, M.S. Zahniser, D.D. Nelson, J.B. McManus, L. Emmenegger, *Appl. Phys. B* **92**, 451 (2008).

[78] C. Bauer, A.K. Sharma, U. Willer, J. Burgmeier, B. Braunschweig, W. Schade, S. Blaser, L. Hvozdara, A. Müller, G. Holl, *Appl. Phys. B* **92**, 327 (2008).

[79] J. Hildenbrand, J. Herbst, J. Wöllenstein, A. Lambrecht, *Proc. SPIE* **7222**, 72220B (2009).

[80] J. Röpcke, P.B. Davies, S. Glitsch, F. Hempel, N. Lang, M. Nägele, A. Rousseau, S. Wege, S. Welzel, *Proc. SPIE* **7222**, 722205 (2009).

[81] L. Mechold, Ph.D. Thesis, University Greifswald, ISBN 978-3-89722-687-6, 2001.

[82] F. Hempel, Ph.D. Thesis, University Greifswald, ISBN 978-3-8325-0262-1, 2003.

2   Infrared absorption spectroscopy

Relevant aspects of infrared laser absorption spectroscopy (IRLAS) in respect to measurements in low pressure reactive plasmas are presented in this chapter. Focussing on semiconductor based lasers, specifically tuneable diode lasers (TDLs) and quantum cascade lasers (QCLs), section 2.1 provides an overview of currently available infrared light sources. The basics of optical absorption theory encompassing linear and non-linear behaviour are collected in section 2.2 and different approaches of sensitivity enhancement are discussed (section 2.3).

2.1   Mid infrared light sources

2.1.1   Overview

Chemical sensing in complex gas mixtures, among them reactive plasmas, generally requires diagnostic techniques of high sensitivity and selectivity. Both criteria may be fulfilled by high resolution (molecular) absorption spectroscopy. The mid infrared (MIR) spectral range (3 ... 20 µm), where strong fundamental transitions of molecular species are located, is typically preferred. An ideal compact light source for absorption spectroscopy should provide continuously tuneable radiation of narrow spectral width and relatively high output power (i.e., high spectral brightness) enabling a selective identification of gas phase constituents and their quantification at a high signal-to-noise ratio (SNR) to be carried out. Since such a source still does not exist, a wide variety of light sources has been used so far and are briefly discussed below. In addition, emerging semiconductor based alternatives, i.e. QCLs and their totally different operation principle compared to conventional TDLs, are considered.

The focus of this section is on narrow band width MIR sources. Hence, techniques that are based on continuous light sources, i.e. dispersive spectrometers and Michelson interferometers comprising Fourier transform IR (FT-IR) spectrometers, are not included here. Nevertheless, FT-IR spectrometer resolution may be as high as 0.001 cm^{-1} but at the expense of recording time and spectrometer size [1,2]. For a similar reason free electron lasers (FELs) providing intense laser pulses also in the MIR (e.g., 5 ... 110 µm, [3,4]) region are neglected, because this radiation is only available at large facilities.

An overview of potential IR sources and related materials is given in figure 2.1 and the appendix (A.1) following the classification of Tittel et al. [5]. They distinguish between direct generation of laser radiation and non-linear frequency down conversion of mainly near infrared (NIR, 0.8 ... 3 µm) (pump) sources, respectively. Apart from Tittel's review more details can be found elsewhere and in the wealth of references therein [6-10].

Line tuneable gas lasers have been used for spectroscopic purposes for many years, especially as long as the performance of semiconductor diode lasers in the MIR was not sufficient. Watt-level output power is easily achieved either in pulsed or continuous wave (cw) mode. However, available wavelengths are limited to pressure broadened ro-vibrational laser transitions at 10/11 µm (CO_2 and isotopes), 5 - 8 µm (CO) and 3 - 5 µm (HF, HCl, HBr, N_2O, CO overtones) respectively [7].

Another important class of continuously tuneable direct MIR sources is based on vibronic transitions occurring in doped crystalline and fibre lasers. Typical dopants are transition-metal (Co^{2+}, V^{2+}) or rare-earth ions (Tm^{3+}, Ho^{3+}, Er^{3+}) [6,8]. Of special interest for the MIR spectral range are Cr^{2+} and Fe^{2+} doped II-VI semiconductors (ZnSe, ZnS) where the laser transition is located in the band gap [8]. Similarly, laser transitions of colour centres in alkali halides (RbCl:Li, KCl:Tl) produce short wavelength MIR radiation [6]. Common to all lasers throughout this class is the requirement of optical pumping by means of adequately powerful sources resulting in output laser powers between tens of mW up to several W. Room temperature and cw operation is now commonly achieved.

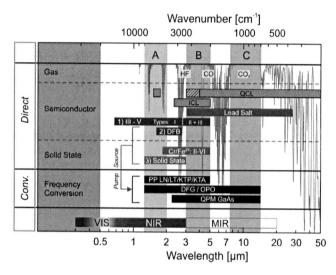

Figure 2.1:   Wavelength coverage of typical infrared (laser) light sources. Radiation might be obtained from direct sources or by frequency conversion. Absorption features of atmospheric constituents (H_2O, CO_2; ~ 1 m absorption path, ambient conditions) are indicated (grey). The overtone region (A) and both atmospheric windows (B, C) are important for chemical sensing applications. (1) diode lasers (binary, ternary, etc. III-V compounds) of type I, II or III (fig. 2.4 a - c), 2) distributed feedback diode lasers, 3) (doped) solid state and fibre lasers/amplifiers. For acronyms see text or glossary. For detailed information on materials see table A.1. Adapted from [5].)

Since the above mentioned solid state and fibre lasers or amplifiers typically cover wavelengths shorter than 4 µm, they are mainly used as pump sources for parametric frequency conversion by means of difference frequency generators (DFGs) and optical parametric oscillators (OPOs) rather than as primary MIR light sources. DFGs convert a pump and an idler beam at different frequencies into a low frequency signal beam in single pass configuration. A resonant (i.e. cavity based) configuration is used for frequency-down converting a single pump laser into two output waves (idler and signal) which is referred to as OPO [5]. Both methods are based on non-linear materials and tuning is facilitated by either tuning the pump source or temperature tuning of the non-linear crystal. Since energy and momentum, i.e. frequency and phase, of the involved waves have to be conserved (phase matching condition), which is difficult to achieve with birefringent crystals in a preferred collinear beam configuration, quasi phase matching (QPM) has been introduced. QPM by applying external electric fields to periodic short regions of ferroelectric materials in non-linear crystals is obtained, e.g., in periodically poled (PP) lithium niobate (LN), lithium tantalite (LT), potassium titanyl phosphate (KTP) and potassium titanyl arsenate (KTA) [5,8]. The wavelength range up to 5 μm is covered by these materials. In order to generate longer wavelengths uncommon materials and pump sources are required [8]. The development of non-linear frequency mixing in GaAs thin films is progressing [11,12]. DFGs typically provide μW ... mW output power due to their limited conversion efficiency whereas tens of mW are observed - still often limited to pulsed mode - by using OPOs in combination with PP materials. Since cw OPOs are increasingly available they may become a competitive alternative MIR source in the spectral range below 5 μm [8]. Nevertheless, the requirement of additional pump sources and sophisticated optical geometries only for generating of radiation remain drawbacks of this approach.

Finally, relatively compact semiconductor based lasers represent an emerging group of direct sources generating light due to stimulated emission across the band gap or quantised energy levels. While diode lasers provide single mode cw operation at room temperature with tens of mW output powers in the NIR range [5], their MIR counterparts, i.e. lead salt lasers, normally require cryogenic cooling for obtaining less than a few mW multimode radiation. The recent advent of cascaded laser structures has made tens and hundreds of mW single mode MIR radiation possible. The basics of light generation in diode and cascade lasers and the clearly different device properties arising from these methods of achieving laser emission are summarised in the subsequent sub-sections.

2.1.2 Tuneable diode lasers

Diode lasers consist of semiconductor alloys forming a p-n junction where population inversion is accomplished by applying a forward bias to the device (figure 2.2). The laser crystal is shaped into a (Fabry-Perot) optical cavity of ~ 0.5 mm length. Hence the multiple cavity modes are typically separated by several cm^{-1} (~ 100 GHz). The energy and thus the emitted wavelength is determined by the band gap energy of the semiconductor alloy. Lasing requires both electrons injected into the conduction band and holes from the valence band. The basic principle shown in figure 2.2 is referred to as homostructure diode laser. Further development aimed at achieving a better population inversion at room temperature and led to double heterostructure lasers (figure 2.3) comprising binary and ternary alloys deposited as μm thick layers. These structures exhibit alternating band gaps and refractive indices resulting in better waveguiding of the emitted radiation. The formed potential wells serve as to better confine the charge carriers (figure 2.3).

Reducing the layer thickness down to nm leads to quantum wells (QWs) with discrete energy levels. Forthcoming progress in deposition technology being capable of producing multi quantum well (MQW) structures and band gap engineering has led to the development of high performance diode lasers, mainly based on III-V compounds (i.e. GaAs, InP, see appendix A.1). These devices emit from the visible to the NIR spectral range at room temperature. Generally, three types of interband transitions across the band gap in MQWs are distinguished (figure 2.4 a - c) representing a compromise on band gap discontinuity (ΔE_{disc}), sufficient charge carrier confinement, overlap of wavefunctions and sufficient population inversion [10,13]. Unfortunately, Auger recombination being the main loss mechanism at longer wavelengths degrades the performance of III-V based devices drastically. Consequently, MIR cw room temperature operation cannot routinely be achieved. Although emission up to 5 μm is theoretically feasible with III-V devices due to the band gap separation, their cryogenically cooled IV-VI counterparts, known as lead salt lasers, become superior in the short wavelength range of the MIR and are the only alternative between 5 - 30 μm (appendix A.1) [9,10].

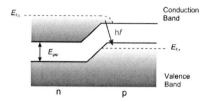
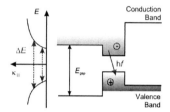

Figure 2.2: Principle of light generation (hf) in a homostructure p-n-junction diode laser of band gap E_{gap} under forward bias. Fermi levels of conduction ($E_{F,c}$) and valence ($E_{F,c}$) band respectively are indicated.

Figure 2.3: Schematic band diagram of a heterostructure diode laser composed of materials with different band gap energies E_{gap} (right). Increased laser line widths ($\Delta v \sim \Delta E$) are caused by the curvature of the dispersion relation $E(\kappa_{\parallel})$ of both bands (left).

Spectral tuning of diode lasers is accomplished by a temperature or charge carrier induced change of the refractive index of the gain medium [14,15]. In other words, wavelength tuning over typically 100 cm^{-1} in the case of lead salt lasers is obtained by setting an appropriate combination of operating temperature and injection current. The spectral coverage is not continuous and encompasses mode overlapping, mode hops or spectral gaps. Current tuning also comprises rapid frequency modulation or frequency locking options [16] which makes diode lasers superior to optically pumped MIR sources (section 2.1.1).

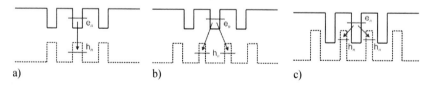

a)   b)   c)

Figure 2.4: Fundamental (radiative) transition between confined electron (e$_n$) and hole (h$_n$) quantum states in (a) type I, (b) type II, and (iii) type III aligned band structures forming multiple quantum wells (after [10]).

Further details on diode laser properties, applications and advances in band gap engineering can be found in [5,7,9,10,15,16] and in their extensive lists of references. Unfortunately, recent efforts in diode laser technology concentrated on telecom wavelengths (~ 1.5 μm), i.e. the spectral region of weaker molecular overtone transitions. The extension of MQW structures to IV-VI lasers having potential of near room temperature operation [e.g., 17,18] or the development of alternative IV-VI compounds [19] were limited by preferential activities on cascaded III-V structures (see next section). Only a few reviews [15,20,21] are concerned with achievements in lead salt laser technology beyond 1990.

2.1.3   Quantum cascade lasers

Similar to MQW the radiative transition in QCLs is based on confined quantum states in stacks of nm thick semiconductor layers. However, in contrast to diode lasers, where electrons and holes are involved, QCLs are unipolar devices, i.e. only electrons travelling between quantum states in the conduction band lead to laser emission (figure 2.5). The energy of the radiation is determined by the separation of the energy levels and thus by the layer thickness providing a means of custom-tailoring the emission wavelength. In contrast to diode lasers, upper and lower laser level have now the same curvature of the dispersion relation (figure 2.3 and 2.6) yielding inherently narrower spectral widths of the transitions [22].

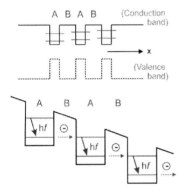
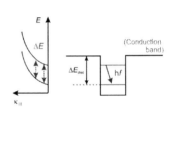

Figure 2.5: **Upper:** Band diagram of a multiple quantum well (MQW) structure without external bias. Confined quantum states of the active zone (A) and the tunneling barrier (B) are schematically shown. Space coordinate is symbolised by x. **Lower:** Radiative transitions (hf) and resonant tunneling between cascaded active (A) and injector (B) regions in a biased and therefore aligned MQW structure.

Figure 2.6: Radiative transition (hf) in an unipolar QCL. Reduced laser line widths follow from the same curvature of the dispersion relation $E(\kappa_{\parallel})$ of both quantum states (left). The band gap discontinuity (ΔE_{disc}) is also schematically shown (right).

Electron transport is accomplished by applying an external electric field to a periodic structure of QWs. Electrons are injected through specifically designed injector regions into the upper laser level followed by the radiative transition into the lower level. It depopulates by tunneling through the potential well barrier which simultaneously serves as the injector of the next active region (figure 2.5). A high tunneling probability is observed for the resonant case, i.e. subsequent lower and upper laser levels are aligned by the external field. Resonant tunneling accompanied by a negative differential resistance of the structure which could be used for light amplification was theoretically predicted in the early 1970s by Esaki [23], Kazarinov and Suris [24], but forthcoming progress in depositing such sophisticated thin film heterostructures was necessary to realise first devices in the late 1980s [25,26]. The potential of a new type of (mid) infrared light source based on linearly polarised intersubband emission in cascaded semiconductor superlattices (SL) by sequential resonant tunneling was recognised [26-28] and introduced by Faist et al. in 1994 as QCL [29,30]. Due to the cascaded structure of typically ~ 30 active regions a single electron is much more often used than in conventional diode lasers yielding quantum efficiencies beyond 100 % and considerably higher output powers compared to lead salt lasers in the range between tens and hundreds of mW.

An efficient population inversion in QCLs requires optimised injector and active region designs (see below) as well as selective depopulation mainly achieved by fast (non-radiative) phonon transitions. Consequently, the wall-plug efficiency, i.e. the portion of input power that is converted into radiation, is rather low and is subject of current improvement [31,32]. The electrical input powers of QCLs are several W due to the relatively high compliance voltages compared to TDLs which are necessary for aligning the involved energy levels. Since typical wall-plug efficiencies are well below 30 % QCLs are thus mainly producing heat. This also hampered cw lasing at room temperature for a long time and required pulsed operation for commercially available devices.

The picture of active zone and injector region in figure 2.5 is oversimplified. Generally, both regions exhibit an intrinsic coupled QW or SL structure [29]. Radiative transitions may take place between coupled quantum states (intersubband) or adjacent minibands (interminiband) rather than single quantum states. Since the first realisation of a QCL several improvements in the structural design have been reported, among them coupled QWs (1-, 2-, 3- or 4-QW [29,33-35]), optical phonon resonance (single up to triple resonance [29,35,36]), bound-to-continuum transitions [37], chirped and graded SLs (i.e., field free SLs by means of varying the SL period or by doping [38-41]), diagonal transitions [42], leaky quantum wells [43], injectorless structures [44] or an optimised number of cascaded stages [45,46]. All proposed designs aimed at facilitating a high population inversion by rapidly depleting the lower level (optical phonon resonance, bound-to-continuum, SLs in active zone) and by simultaneously reducing the internal loss and leakage channels (diagonal transitions, chirped SL with less doping leading to reduced free carrier absorption). This has finally led to room temperature [47] and cw operation [48] of QCLs. Particularly chirped SLs exhibit several advantages for long wavelength devices (beyond ~ 10 µm) [49] and are capable of carrying higher currents which in turn enables higher output powers to be achieved [46]. An extensive review by Gmachl et al. concerns further aspects of QCL development until 2000 [50]. Additional details can be found in similar summaries [46,51-53]. Recently the thermal management, i.e. heat removal from the active zone, and thus the performance of QCLs has been considerably improved by means of buried heterostructures [54], additional barrier layers [55] and alternative laser chip mounting [56] yielding W-level cw output power at room temperature [57].

The majority of QCLs consists of III-V compounds (see appendix A.1). The first QCL comprised GaInAs/AlInAs alloys lattice matched on InP [29]. Soon after GaAs/AlGaAs device were demonstrated [58,59]. In contrast to diode lasers, where the identification of suitable narrow band gap semiconductor alloys for achieving laser emission beyond 5 µm is challenging, tailoring the emission of QCLs to shorter wavelengths is unsatisfactorily solved. Laser emission from GaAs/AlGaAs devices, which are often used for generating THz radiation, cannot fall below ~ 8 µm, while 3.4 µm emission has been reported for strain-compensated GaInAs/AlInAs [60]. Consequently, strong C-H absorption features around

3 μm are not accessible. The reason is the limited band gap discontinuity, ΔE_{disc}, of the employed semiconductor compounds (figure 2.5). Shorter wavelengths require an increased subband or miniband separation which has to be clearly smaller than the band gap discontinuity to avoid parasitic leakage from the upper laser level. Short wavelength QCLs can therefore only be achieved with alloys of similar lattice constants but larger band gap differences than the above mentioned compounds [61]. InAs/AlSb based structures and admixtures of Sb to existing material systems were recognised as potential alternatives [62,63], but commercial availability is still absent. Additionally quaternary or even pentanary systems are investigated (see appendix A.1 for references) and lasing at optical communication wavelength of 1.55 μm was reported for GaN/AlGaN [64]. Independent from material engineering activities the combination of cascaded structures and type III interband transitions (figure 2.4 c), known as interband cascade lasers (ICLs), was proposed [43,65-68]. In this case high quantum efficiencies for short wavelengths are obtained by using bipolar devices and transitions across staggered quantum wells [69-74].

Similar to diode lasers a QCL chip forms a Fabry-Perot cavity of a few mm length resulting in a multimode emission spectrum of ~ 1 cm^{-1} mode spacing. Tuneable single mode operation is achieved (i) in external cavity (EC) configuration or (ii) by means of distributed feedback (DFB).

Although an EC arrangement employing uncoated multimode pulsed QCLs has been successfully demonstrated [75], anti-reflection coated lasers are desirable for efficient spectral narrowing [76-78]. Temperature tuning as well as tilting the external grating yielded a spectral coverage of up to 20 % around the centre wavelength for a pulsed system [79] and a ~ 8 % mode hop free tuning range for a cw system [80] which are now also routinely available parallel to the earlier commercialised DFB QCLs.

Particularly the highly developed processing technology for the initially employed III-V QCL compounds enabled DFB gratings to be integrated into the QCL structure, more specifically in the waveguide structure [81]. Bragg reflection in complex- [81,82] or index-coupled [83] gratings provide a means of observing single mode QCL emission by selecting one of the cavity modes. Spectral tuning is accomplished by temperature induced changes of the refractive index of the laser which tunes both the spectral gain and to a lesser extent the period of the Bragg grating [50]. Hence, the QCL current can only be used indirectly for sweeping the laser wavelength through internal laser heating. The total emission range of DFB QCLs is typically limited to less than 7 cm^{-1} between - 30 °C and + 30 °C. Additionally, Faist et al. found an inherent thermal drift in pulsed DFB QCLs caused by the dissipated power in the active region which is also known as chirped pulse [81,84].

The previously discussed issues, especially the wavelength chirp of pulsed QCLs, strongly indicates that QCLs are far more than a substitute to TDLs [85], e.g., lead salt lasers, and the application of state-of-the art TDL methods to QCLs have to be carefully validated (chapter 4).

2.2   Optical absorption theory

Essential equations for both linear and non-linear optical absorption phenomena which are necessary for analysing the experimental data in the subsequent chapters are provided in this section. Since measurements in the infrared spectral range and, particularly, the database values for molecular line parameters are commonly expressed in wavenumbers the corresponding equations are often given in or adapted to cgs units [86-90 and references therein]. Although this thesis is generally based on SI units, the main expressions in this section refer to the *HITRAN* notation. Exceptions from the SI standard are mentioned and a set of SI based equations, e.g., used in [91,92], can be found in the appendix A.3.

Note that the symbol $\nu = 1/\lambda$ (in cm^{-1}) denotes wavenumbers throughout the thesis while $f = c\nu$ (in s^{-1} or Hz) represents frequencies and c is the velocity of light. The angular frequency $\omega = 2\pi\nu$ is not employed. Additionally, the number of molecules N per unit volume (or number density) is symbolised by $n = N/V$. This is sometimes referred to as concentration. However, in this thesis the terms mixing ratio and concentration are used synonymously for the relative value n/n_{tot} where n_{tot} follows from the ideal gas law, i.e. the total pressure p_{tot}, the gas temperature T and the Boltzmann constant k_B

$$p_{tot} = n_{tot}k_B T .$$
(2 - 1)

2.2.1   Linear absorption

Considering the transmission of incident radiation $I_{in}(\nu)$ in a one dimensional gas sample, the variation of the intensity, $dI(\nu)$, after passing through a differential slab dz is described by the radiative transfer equation [86] (figure 2.7)

$$dI(\nu) = (j(\nu) - k(\nu))I_{in}(\nu)dz ,$$
(2 - 2)

where $j(\nu)$ is the emission coefficient and $k(\nu)$ is the absorption coefficient (in cm^{-1}). In this case only linear effects are accounted for ($dI \sim I_{in}$). The absorption coefficient $k(\nu)$ comprises contributions from all absorption features, indexed by i, at wavenumber ν, with their absorption cross section $\sigma_i(\nu)$ and their corresponding number density of species n_i per unit volume:

$$k(\nu) = \sum_i k_i(\nu) = \sum_i n_i\sigma_i(\nu) .$$
(2 - 3)

Due to the strong ν^4 dependence scattering can normally be neglected in the infrared compared to the visible spectral range. Assuming a homogeneous sample containing only one species where no emission and scattering are present, equation (2 - 2) simplifies to $dI(\nu) = -k(\nu)I_{in}(\nu)dz$ and integration over an effective absorption path L_{eff} leads to the Beer-Lambert law of linear absorption

$$\ln\left(\frac{I_0(\nu)}{I(\nu)}\right) = k(\nu)L_{eff} .$$
(2 - 4)

The effective length L_{eff} encompasses all absorption techniques, i.e. single-pass absorption, folded optical paths (multiple pass absorption) or the transmission through optical cavities[1]. $I_0(\nu)$ and $I(\nu)$ denote the incident and the transmitted radiation through the sample, (i.e. $I_0 = I(z = 0)$ and $I = I(z = L_{eff})$, figure 2.7 and 2.8). The intensity I (in W/m^2) of a light wave is connected to the electrical field amplitude E_0 by (in SI units) [91,92]

$$I = \frac{1}{2} c \varepsilon_0 E_0^2,$$ (2 - 5)

where ε_0 denotes the dielectric constant.

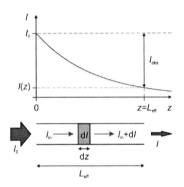

Figure 2.7: 1D-illustration of the radiative transfer equation (2 - 2) and Beer-Lambert law (eq. (2 - 4)).

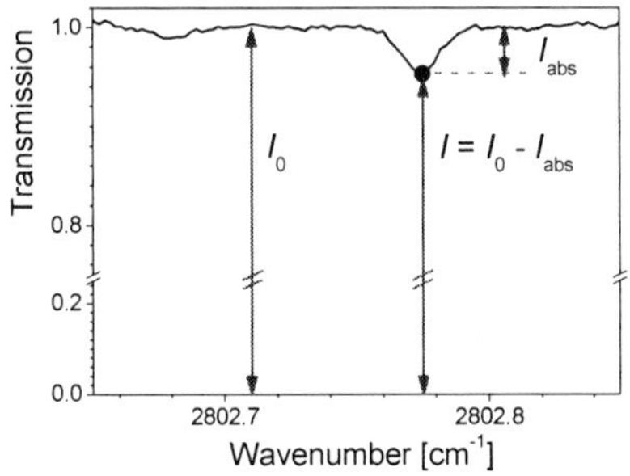

Figure 2.8: Transmission spectrum of a H$_2$CO absorption feature (observed in an Ar/CH$_4$/N$_2$/O$_2$ plasma) employing the symbols defined in figure 2.7.

Using equation (2 - 4) the integrated absorption coefficient K (in cm^{-2}) of an isolated absorption line provides a means of calculating the absolute number density n from a relative measurement and is defined as

$$K = \int_{-\infty}^{+\infty} k(\nu) d\nu = \frac{1}{L_{eff}} \int_{-\infty}^{+\infty} \ln\left(\frac{I_0(\nu)}{I(\nu)}\right) d\nu = nS \ .$$ (2 - 6)

Thus, the line strength S (in cm/molecule) may be considered as proportionality factor to describe the absorption cross section σ (in cm^2) of a transition centred at ν_0 (cf. eq. (2 - 3))

$$\sigma(\nu - \nu_0) = S \cdot \phi(\nu - \nu_0),$$ (2 - 7)

with $\phi(\nu - \nu_0)$ being the normalised line profile (for examples see the appendix A.2)

$$\int_{-\infty}^{+\infty} \phi(\nu - \nu_0) d\nu = 1.$$ (2 - 8)

---

[1]   A collection of formulae concerning absorption spectroscopy using optically resonant cavities is given in 6.2.

2.2.2   Line profiles and spectral broadening

The line shape of an absorption line, as described by the profile $\phi(\nu - \nu_0)$, depends on the experimental conditions, among them gas temperature T and pressure p, and is often characterised by its half width at half maximum (HWHM) or spectral width $\Delta f = c \cdot \Delta \nu$. The spectral width Δf may also be expressed in terms of an equivalent transition rate γ or time τ

$$\gamma = \frac{1}{\tau} = 2\pi \cdot \Delta f \,. \tag{2-9}$$

An absorption feature can be affected by
   i)   natural line broadening,
   ii)   transit time broadening,
   iii)   collisional broadening, and
   iv)   inhomogeneous line broadening.
While the inhomogeneous Doppler broadening iv) results in a Gaussian line profile, the homogeneous effects i) - iii) cause Lorentzian line shapes. The natural line width i) is determined by the spontaneous life time being in the order of $\tau_{sp} = 10^{-3}$ s for vibrational levels in the electronic ground state of molecules. Thus, the natural spectral broadening of infrared transitions is typically smaller than 1 kHz $(3 \times 10^{-8}$ cm$^{-1})$ [92,93]. Depopulation of excited levels may result from collisions between molecules. The collision frequency τ_{col}^{-1} is described by the collision cross section, the number density of molecules (eq. (2 - 1)) and the mean thermal velocity ν_{th}

$$\nu_{th} = \sqrt{\frac{8 k_B T}{\pi m_{molec}}} \,, \tag{2-10}$$

where $m_{molec} = M_{mol}/N_A$ is the molecular mass which can also be expressed by the molar mass M_{mol} and the Avogadro number N_A. Hence, the spectral width due to collisional broadening iii) is pressure and temperature dependent and described by

$$\Delta f_{col} \sim \gamma_p \frac{p}{\sqrt{T}} \,. \tag{2-11}$$

This is often referred to as pressure broadening where the pressure broadening coefficient γ_p (HWHM) is tabulated (section 2.2.3)[2]. Molecular ro-vibrational transitions exhibit pressure broadened half widths of a few MHz/mbar [91-93], i.e. ≤ 3 MHz $(0.0001$ cm$^{-1})$ at low pressures < 5 mbar and ~ 1.5 GHz $(0.05$ cm$^{-1})$ at atmospheric pressure.
Doppler broadening iv) due to the thermal motion of molecules yields a spectral width (HWHM) of [91,92]

$$\Delta f_D = f_0 \sqrt{\frac{2 k_B T \ln 2}{m_{molec} c^2}} = 3.58 \cdot 10^{-7} \sqrt{\frac{T}{M_{mol}}} f_0 \,. \tag{2-12}$$

---

[2]   The collision frequency is sometimes defined as $2/\tau_{col}$ [93]. Since γ_p is often taken from databases and given by an empirical formula similar to (2 - 11) a detailed discussion is omitted here.

The right hand side of equation (2 - 12) requires the temperature and the molar mass to be given in K and $g \cdot mol^{-1}$, respectively. Doppler widths of 30 ... 90 MHz (0.001 ... 0.003 cm^{-1}) at room temperature are typically observed for molecular transitions in the infrared [91-93]. Doppler broadening therefore clearly exceeds the pressure broadened line widths under the low pressure conditions.

Due to the narrow natural line widths in the infrared spectral range transit time broadening ii) may be considered, i.e. the limited interaction time $\tau_{tt} = \Delta x_{beam}/v_{th}$ of the molecules with the incident laser beam. The corresponding spectral width is

$$\Delta f_{tt} = \frac{1}{2\pi} \frac{v_{th}}{w} \sqrt{2 \ln 2} \qquad (2 - 13)$$

in the case of a parallel Gaussian beam with diameter $2w$ [91,92]. Hence, at room temperature ($v_{th} \approx 500$ m/s) transit time broadening for a laser beam of 5 mm diameter (19 kHz, 6×10^{-7} cm^{-1}) exceeds the natural line width.

The superposition of the homogeneous contributions for ro-vibrational transitions with

$$\tau_{sp} > \tau_{tt} \gg \tau_{col} \qquad (2 - 14)$$

yields a mainly pressure broadened Lorentzian profile. At elevated and atmospheric pressure (> 50 mbar) the homogeneous spectral width exceeds the inhomogeneous half width and hence the line shape can be assumed to be of Lorentzian type which can be described analytically. This provides a means to estimate the integrated absorption coefficient K (or the molecular number density) using only the peak absorption at centre wavenumber v_0 and the Lorentzian half width $\Delta v_{col} = \Delta f_{col}/c$ (in cm^{-1} to fulfil eqs. (2 - 6), (2 - 7) and (2 - 11)) without integrating the line:

$$K = nS = \frac{\pi}{L_{eff}} \ln \left(\frac{I_0(v)}{I(v)} \right) \Bigg|_{v_0} \Delta v_{col} . \qquad (2 - 15)$$

Similarly, under low pressure conditions the line shape may be approximated by a Gaussian profile with the Doppler half width $\Delta v_D = \Delta f_D/c$ from eq. (2 - 12) resulting in

$$K = nS = \frac{1}{L_{eff}} \sqrt{\frac{\pi}{\ln 2}} \ln \left(\frac{I_0(v)}{I(v)} \right) \Bigg|_{v_0} \Delta v_D . \qquad (2 - 16)$$

For the intermediate pressure range a convolution of both line profile types, i.e. a Voigt profile, is observed which typically requires a numerical calculation. In this case K should be obtained by integrating over the absorption feature. Equations (2 - 15) and (2 - 16) also enable the minimum detectable number density for a selected transition to be estimated if the peak-to-peak noise is considered as peak absorption.

In the previous discussion the instrumental broadening Δv_{instr} was assumed to be substantially smaller than the main broadening contribution. Particularly at low pressure conditions the laser line width of lead salt lasers may be of the same order as the Doppler broadening. The convolution of absorption and instrumental profile, which is assumed to be Gaussian as well, yields the observed spectral width, $\Delta \nu_{obs} = (\Delta \nu_D^2 + \Delta \nu_{instr}^2)^{1/2}$, which replaces $\Delta \nu_D$ in equation (2 - 16).

2.2.3   Infrared absorption cross sections and databases

The knowledge of (infrared) absorption cross sections is essential for both the identification of molecular species in acquired spectra and for the retrieval of quantitative concentration values. Different types of gas phase spectroscopic databases are available in the literature. The most convenient case for practical purposes is a high resolution "line-by-line" record where molecular line parameters such as S, γ_p or individual temperature dependences are listed for each transition which would enable a direct determination of an unknown mixing ratio according to equation (2 - 6).

Although the number of species and the spectral coverage of high resolution compilations have been continuously extended and the data accuracy has been improved in the past years, alternative approaches still have to be considered. Provided that spectroscopic data of medium resolution (e.g., 0.1 cm^{-1}) are available, the integrated band intensity may be used to derive line strengths for all transitions of this band [94]. This, however, still requires predictions for individual transition parameters which could be calculated for simple molecule geometries. Nevertheless, spectral reference libraries based on FT-IR studies, as provided by the National Institute of Standards and Technology (NIST, 0.13 cm^{-1} resolution, 40 molecules) or the Pacific Northwest National Laboratory (PNNL, 0.11 cm^{-1} resolution, 280 molecules) [95,96], may help identifying especially complex absorption features of larger molecules or calibrating the own instrument. Unfortunately both quantitative infrared databases provide pressure broadened spectra at ambient conditions which may clearly be different from low pressure plasma conditions with increased gas temperatures. If reference spectra are absent, a customised calibration is feasible for stable species but difficult to realise for transient molecules. In some cases absolute values can be obtained by using a kinetic decay method if the main removal process is known [e.g., 97].

In respect to the present work high resolution compilations are of special interest. Their history and (future) development has been reviewed in [98]. A few aspects concerning their application for low pressure plasma diagnostics are discussed in what follows.

The most common molecular spectroscopic databases focussing on atmospheric constituents are known as *HITRAN* (*HI*gh resolution *TRAN*smission) and *GEISA* (*G*estion et *E*tude des *I*nformation *S*pectroscopiques *A*tmosphérique) initially published in 1973 and 1976 [99,100], respectively, and are periodically updated [90,101-110]. Other catalogues have been compiled in the past decades concentrating on the microwave and millimetre spectral range [111-113], astronomic applications [111,114,115] or satellite experiments by selecting specific molecules (e.g., *ATMOS* [116-119]). With a few exceptions *HITRAN* and *GEISA* contain the same species (~ 40) in their line-by-line part. The *GEISA* data format was adapted to the *HITRAN* notation (section 2.2.4) and is therefore similar or may easily be converted. Typically the spectroscopic parameters are tabulated for a reference temperature $T_{ref} = 296$ K [99,108]. Line strength values, as defined by equation (2 - 6), are normalised to the molecular number density n. Several other notations are reported (e.g., normalised to the pressure, using cgs, SI or customised units, etc.). A summary is given in [88].

The increasing use of the compilations for non-atmospheric applications with temperatures different from ambient conditions, e.g., in astronomy or for plasma diagnostics, has led to substantial changes in the early format of the fundamental line parameters [120]. While the line strength includes the isotopic abundance I_a, the transition dipole moment μ is sometimes desirable (section 2.2.5). Hence, *GEISA* provides also μ in the current version [109]. The *HITRAN* compilation encompasses now the Einstein coefficient A [106]. Additionally, the temperature dependence of the total internal partition function $Q(T)$ is supplied along with *HITRAN* [121-123], which is necessary to correct the tabulated line strength $S(T_{ref})$ into the experimentally relevant $S(T \neq T_{ref})$. Earlier versions used a polynomial fit to $Q(T)$ and provided only the corresponding coefficients. In recent editions (i.e. after 2003) $Q(T)$ is tabulated for each isotope between 70 and 3000 K.

The previous discussion clearly suggests the use of the *HITRAN* database for plasma diagnostic purposes since the required spectroscopic parameters for main (stable) molecules have been carefully established for a long time and correction terms for deviations from the reference temperature are included. For this reason quantitative results in this thesis are based on the *HITRAN* compilation including the cgs related notation. Relevant equations to convert the listed parameters are provided in the next section.

2.2.4   Line strength and molecular line parameters

Considering an (open) two-level system [91,92] with a (*l*)ower and an (*u*)pper level and following [86,87,94,120] the line strength S_{lu} for a transition $l \rightarrow u$ normalised to the total number density n is expressed (in cgs units) by

$$S = S_{lu} = \frac{8\pi^3}{3hc} \nu_{lu} \frac{n_l}{n} \cdot \left[1 - \frac{g_l n_u}{g_u n_l} \right] \cdot \left\{ \frac{1}{g_l} |\mu_{lu}|^2 \right\}, \qquad (2 - 17)$$

where n_l, n_u are the number density of molecules in the states l and u with their corresponding statistical weights g_l, g_u, h denotes the Planck constant, $|\mu_{lu}|$ is the transition dipole moment and ν_{lu} (in cm^{-1}) is the frequency of the transition, respectively. The term in braces in equation (2 - 17) is known as weighted transition dipole moment. The line strength or the transition probability depends obviously on the population difference between the lower and the upper level. If the system is in thermal equilibrium at temperature T, the term in square brackets can be rearranged by using the Maxwell-Boltzmann distribution law (SI units)

$$\frac{n_l}{g_l} = \exp\left(- \frac{E_l}{k_B T} \right) \frac{n}{Q(T)} \qquad (2 - 18)$$

with the energy of the lower state E_l and the total internal partition function $Q(T)$. The equation for n_u/g_u is analogue. This leads to a form of S (in cm/molecule) which accords with the notation of the *HITRAN* database

$$S = S_{lu} = \frac{8\pi^3}{3hc} \frac{I_a g_l}{Q(T)} \nu_{lu} \exp\left(- \frac{hcE_l}{k_B T} \right) \left[1 - \exp\left(- \frac{hc\nu_{lu}}{k_B T} \right) \right] \cdot \left\{ \frac{1}{g_l} |\mu_{lu}|^2 \right\}. \qquad (2 - 19)$$

The expression in square brackets describes the emission from the upper level which is generally non-negligible in the infrared spectral range. Using this notation $j(\nu)$ can be considered as zero in equation (2 - 2), because $k(\nu)$ already accounts for the emission. Note that the squared weighted transition dipole moment is often provided in (Debye^2) which requires an additional factor 10^{-36} [89,90,120] and that E_l is now in cm^{-1} leading to the factor (hc) in the exponent of the Boltzmann term.

Since the line strength in *HITRAN* is defined for the reference temperature $T_{\text{ref}} = 296$ K, $S(T)$ must be corrected by using the tabulated values $S(T_{\text{ref}})$, E_l, ν_{lu} and $Q(T)$ [88,90]

$$S(T) = S(T_{\text{ref}}) \frac{Q(T_{\text{ref}})}{Q(T)} \frac{\exp(-hcE_l/k_BT)}{\exp(-hcE_l/k_BT_{\text{ref}})} \cdot \frac{[1 - \exp(-hc\nu_{lu}/k_BT)]}{[1 - \exp(-hc\nu_{lu}/k_BT_{\text{ref}})]}. \qquad (2 - 20)$$

The line strength in (2 - 19) and (2 - 20) is not defined for a pure gas, but for an isotopic abundance I_a. However, in some cases, e.g., for non-linear phenomena, the transition probability A_{ul} or dipole moment $|\mu_{lu}|$ is the relevant parameter (section 2.2.5). A small I_a, being typically less than a few percent for minor isotopes [88], results in a weak line strength, but the dipole moment for the isotopic line may still be high. Therefore the conversion between S_{lu}, μ_{lu} and the Einstein coefficients A_{ul}, B_{lu} and B_{lu} is important.

The Einstein coefficients B_{lu} and B_{ul} of induced absorption and emission obey

$$g_l B_{lu} = g_u B_{ul}, \qquad (2 - 21)$$

where B_{lu} is connected with the transition dipole moment (in cgs units) by [90,120]

$$B_{lu} = \frac{8\pi^3}{3h^2} \left\{ \frac{1}{g_l} | \mu_{lu} |^2 \right\}. \qquad (2 - 22)$$

The Einstein coefficient of spontaneous emission A_{ul} can be expressed in terms of B_{lu} and hence also by μ_{lu} (in cgs units) [90,120]

$$A_{ul} = 8\pi h \nu_{lu}^3 \cdot \frac{g_l}{g_u} B_{lu} = \frac{64\pi^4}{3h} \nu_{lu}^3 \left\{ \frac{1}{g_u} | \mu_{lu} |^2 \right\}. \qquad (2 - 23)$$

Note that ν_{lu} (in cm^{-1}) still denotes wavenumbers. Transforming (2 - 19) yields [120]

$$| \mu_{lu} |^2 = S_{lu} \frac{3hc}{8\pi^3} \cdot \frac{Q(T)}{\nu_{lu} I_a} \cdot \frac{1}{\exp(-hcE_l/k_BT)[1 - \exp(-hc\nu_{lu}/k_BT)]} \qquad (2 - 24)$$

and inserting into (2 - 23) [88,120]

$$A_{ul} = 8\pi \nu_{lu}^2 \cdot c \cdot \frac{Q(T)}{\exp(-hcE_l/k_BT)[1 - \exp(-hc\nu_{lu}/k_BT)]} \cdot \frac{1}{I_a g_u} S_{lu}. \qquad (2 - 25)$$

In contrast to the line strength S_{lu}, both values μ_{lu} and A_{ul} are independent from the isotopic abundance. While the weighted transition dipole moment was given additionally in earlier *HITRAN* editions, now A_{ul} is supplied enabling μ_{lu} to be calculated by means of (2 - 23). A comprehensive summary and more details on the different relations in respect to the *HITRAN* notation can be found in [88,120] and in the appendix of [90]. A set of equations in SI units is provided in the appendix A.3.

As mentioned in the previous section, the contribution of collisional broadening to the spectral line width is pressure and temperature dependent and usually described by a pressure broadening coefficient (eq. (2 - 11)). *HITRAN* lists two values (HWHM), i.e. the air broadening coefficient $\gamma_{air,ref}$ and the self broadening coefficient $\gamma_{self,ref}$ (in cm^{-1}/atm), respectively, at the reference conditions ($T_{ref} = 296$ K and $p_{ref} = 1$ atm (1013.25 mbar)). The corresponding pressure broadened half width is then

$$\frac{\Delta f_{col}}{c} = \left(\frac{T_{ref}}{T}\right)^m \left(\gamma_{air,ref}\left(p_{tot} - p_s\right) + \gamma_{self,ref}\, p_s\right) \tag{2 - 26}$$

with total pressure p_{tot} and partial pressure p_s (in atm). Equation (2 - 11) suggests an exponent $m = 0.5$. In some case different values might be provided. For trace gas measurements or experiments at low pressure conditions the self broadening contribution is typically negligible.

2.2.5   Non-linear absorption

It is clear from the previous discussion that the line strength S_{lu} and thus (via eqs. (2 - 7) and (2 - 3)) the absorption coefficient $k(\nu)$ is proportional to the population difference Δn between the upper and the lower level ($S_{lu} \sim (n_l/g_l - n_u/g_u)$, eq. (2 - 17)). However, in the presence of strong radiation fields the population of the upper level is not determined by thermal equilibrium (Boltzmann law). The population difference is reduced and becomes intensity dependent and so is the absorption coefficient.

Consequently, $dI(\nu) = -k(\nu)I_{in}(\nu)dz$ (eq. (2 - 2)) requires a modification. The phenomenological approach is to introduce a non-linear factor, e.g. $dI(\nu) = -[k(\nu)I_{in}(\nu) - q(\nu)I^2_{in}(\nu)]dz$ [91], which might serve as a check on the validity of the Beer-Lambert law: if the absorption does not follow linearly the increasing number density or pressure, corrections have to be considered.

Assuming the (open) two-level system, non-linear absorption behaviour is usually described by a saturation parameter Σ

$$\Sigma = \frac{\Omega_{lu}^2}{\gamma_l \gamma_u} \approx \left(\frac{\Omega_{lu}}{\gamma}\right)^2 \tag{2 - 27}$$

which is generally the ratio of the optical pumping rate to the relaxation rate (γ) of the system [91,92,94]. The pumping rate is described by Ω_{lu}, i.e. the Rabi flopping frequency[3] at the centre frequency ν_{lu}

$$\Omega_{lu} = \frac{\mu_{lu} E_0}{h} \tag{2 - 28}$$

where E_0 is the electric field amplitude following from equation (2 - 5) to $(2I/(c\varepsilon_0))^{1/2}$ [91,92,94]. The relaxation rate in equation (2 - 27) follows from the rates of the upper γ_u and the lower level γ_l, respectively, being comparable to a mean relaxation rate γ in the infrared spectral range [94]. A detailed discussion on this assumption for the microwave region is given in [124] and may still be valid for the mid infrared spectral range.

The equilibrium population difference Δn_{eq} in the (open) two-level system is then reduced to

$$\Delta n(I) = \frac{\Delta n_{eq}}{1 + \Sigma} = \frac{\Delta n_{eq}}{1 + I/I_{sat}} \tag{2 - 29}$$

and becomes intensity dependent [91,92]. The saturation intensity I_{sat} is defined for the case $\Sigma = 1$ and follows from (2 - 5), (2 - 27) and (2 - 28). In order to estimate the consequences for the absorption coefficient $k(\nu)$ the absorption profile has to be considered as well. For a homogeneously broadened transition the absorption coefficient is reduced by a frequency dependent factor

$$k_{hom}^{sat}(\nu) = \frac{k_{hom}^{eq}(\nu)}{1 + \Sigma(\nu)} \tag{2 - 30}$$

where $\Sigma(\nu)$ has its maximum, equal to (2 - 27) at the centre frequency ν_{lu} [91,92]. The line profile remains Lorentzian with an increased half width

$$\Delta f_{sat} = \Delta f_{hom} \sqrt{1 + \Sigma} . \tag{2 - 31}$$

In the case of inhomogeneous broadening the equilibrium absorption coefficient (and hence the line strength S_{lu}) is constantly damped by

$$k_{inhom}^{sat}(\nu) = \frac{k_{inhom}^{eq}(\nu)}{\sqrt{1 + \Sigma}} = \frac{k_{inhom}^{eq}(\nu)}{\sqrt{1 + I/I_{sat}}} . \tag{2 - 32}$$

The homogeneous broadening is also increased as described by (2 - 31) and may be then non-negligible compared to the inhomogeneous half width.

---

[3] Here the Rabi frequency is defined as equivalent to f in Hz instead of an angular frequency.

If relaxation processes are treated more generally than in the previous discussion, two characteristic relaxation times are defined:

$$T_1 = \frac{1}{\gamma_l} + \frac{1}{\gamma_u} = \frac{\gamma_l + \gamma_u}{\gamma_l \gamma_u}, \tag{2-33}$$

$$T_2 = \frac{1}{\gamma_l + \gamma_u}. \tag{2-34}$$

T_1 is the longitudinal relaxation or population decay time and encompasses all depopulation processes (e.g., spontaneous emission, inelastic collisions)[4] while T_2 is known as transversal relaxation or phase decay time. The latter describes the perturbation of the phase relation between the driving electrical field and the induced polarisation of the system [91,92,124-126]. T_2 typically consists of homogeneous and inhomogeneous contributions [127].

Using (2 - 33) and (2 - 34) the saturation parameter from equation (2 - 27) becomes $\Sigma = \Omega^2_{lu} T_1 T_2$ being a common form of expressing coherent (non-linear) phenomena in atoms or molecules. It is adapted to the Bloch vector approach which was originally developed for magnetic resonance experiments [128,129]. It covers both saturation and transient effects. The latter encompasses transient absorption as well as transient emission and fast or rapid passage effects, i.e. the incident field and the corresponding transition are swept out resonance relatively fast [124]. Fast passage is accomplished by sweeping the source or detuning the resonance frequency by means of external fields on time scales shorter than the relaxation times T_1 and T_2. The absorption feature and thus the absorption coefficient exhibits then an oscillatory structure after switching the driving field [124,127]. These "wiggles" have been observed in nuclear magnetic resonance [128,130,131], in electron spin resonance [132,133] and in the field of optical coherent spectroscopy [124,127] comprising the visible [134,135], infrared [136,137] and microwave spectral range [138-141].

Fast passage effects where the incident radiation is swept with a sweep rate α (in Hz/s) are usually described by a normalised sweep rate A [129]

$$A = \alpha T_1 T_2. \tag{2-35}$$

The ratio of the normalised sweep rate and the saturation parameter

$$\frac{A}{\Sigma} = \frac{\alpha}{\Omega^2_{lu}}. \tag{2-36}$$

provides a criterion to discriminate different coherent absorption phenomena. The most relevant cases are summarised in table 2.1 [129]. Available mid infrared radiation sources are expected not to violate the thermal equilibrium of the absorbing species and so Beer-Lambert

---

[4]   If a (closed) two-level system is considered, consisting only of the two levels l and u without further relaxation possibilities, T_1 is defined as the decay time of the upper level [125,126].

law of linear absorption (i) is typically employed in IRLAS (section 2.2.1). Laser sources of adequately high output power or strong absorption features may require accounting for power saturation effects (ii). An additional frequency sweep of the laser, either inherent to the light source or induced, leads to fast passage behaviour [135,142]. The ratio A/Σ determines if a potential population inversion is reduced (iii) or enhanced (iv) which is also known as (optical) adiabatic rapid passage [136,137].

Regime		Conditions	
i)	slow passage - linear regime	$A \ll 1, \Sigma \ll 1$	$A/\Sigma \ll 1$
ii)	slow passage - saturation	$A \ll 1, \Sigma \gg 1$	$A/\Sigma \ll 1$
iii)	fast passage - linear regime	$A \gg 1, \Sigma \gg 1$	$A/\Sigma \gg 1$
iv)	fast passage - adiabatic regime	$A \gg 1, \Sigma \gg 1$	$A/\Sigma \ll 1$

Table 2.1: Relevant non-linear absorption phenomena following the classification of [129].

2.3 Sensitivity considerations

Selecting narrow line width infrared tuneable lasers sources of adequately high spectral output power is essential for achieving high sensitivity and selectivity, respectively. The narrow spectral width provides high selectivity for discriminating absorption features of different molecular species, especially in low pressure samples. Additionally, the absorption scales approximately with the ratio of absorption line width and laser spectral width ($dI \sim \Delta f/\Delta f_{laser}$) [91,92]. Thus, a narrow laser line width a critical parameter for realising a highly sensitive experiment. Further means are briefly classified in what follows.

The sensitivity of an experiment is typically evaluated by the signal-to-noise ratio

$$SNR = \frac{I_{abs}}{\Delta I} = \frac{1-I}{\Delta I} \approx \frac{I_0}{\Delta I} kL_{eff} .$$ (2 - 37)

Transforming the absorption signal I_{abs} into the transmitted intensity, I, through the sample and assuming weak absorption for the exponential form of Beer-Lambert law (eq. (2 - 4)), $I \sim \exp(-kL_{eff}) \approx 1 - kL_{eff}$) yields an estimate of the SNR. Provided that a minimum $SNR = 1$ is achieved, an expression for the minimum detectable absorption coefficient k_{min} or number density (MDND) n_{min} can be deduced from (2 - 37):

$$n_{min}\sigma = k_{min} = \Delta k \approx \frac{\Delta I}{I_0 L_{eff}} .$$ (2 - 38)

Consequently, a higher sensitivity is accomplished by reducing the noise signal ΔI or increasing the absorption path length L_{eff}. Different approaches are briefly discussed below and are limited to direct absorption, e.g., photoacoustic methods are excluded. Typical values are summarised in table 2.2.

In direct absorption without employing additional measures the relative error or minimum detectable fractional absorbance $\Delta I/I_0$ is typically in the order of $10^{-2} \ldots 10^{-4}$. Further reduction is achieved by averaging over several rapidly acquired spectra and harmonic detection schemes respectively [143].

The most common modulation or derivative techniques might be referred to as frequency modulation in general. In practice frequency modulation (FM) spectroscopy applying a modulation signal in the MHz range clearly exceeding the absorption line width is often distinguished from wavelength modulation (WM) spectroscopy where low frequency modulation (kHz range) is used [16]. A recent review is given by Friedrichs [144]. While a minimum fractional absorbance of approximately 10^{-5} has been achieved in a 1 s measurement interval employing WM spectroscopy (2f), FM techniques may decrease the detection limit by additional two orders of magnitude [143,145,146]. Long time averaging over minutes or averaging over thousands of spectra acquired with a kHz sweep rate are alternative and successfully applied methods of reducing $\Delta I/I_0$. This is often accompanied by background subtraction and yields detection limits of $10^{-5} \ldots 10^{-6}$ [143,147,148]. Mechanical fluctuations and optical fringes typically degrade the spectrometer performance leading to a practical lower limit of $\Delta I/I_0 \approx 10^{-6}$ for both averaging or modulation techniques [147,149].

Since the residual noise signal ΔI cannot arbitrarily be reduced the effective absorption path has to be extended to achieve higher sensitivity. Folded optical lines of different geometries where the laser beam is reflected once by a retro-reflector up to several hundred times in sophisticated multi-pass arrangements enable an extension of L_{eff} up to hundreds of metres. Different types of multiple pass configurations are reported in the literature and are based on plane mirrors [150], spherical optics, e.g., White [151], Chernin [152] and Herriott type [153,154] cells respectively, or astigmatic mirrors [155].

The interaction length can further be increased up to the km range by employing optical cavities consisting of high reflectivity mirrors (> 99.9 %) which enclose the sample volume [156-158]. Analysing the radiation leaking out of the cavity and its decay rate yields information on absorptive losses inside and thus on the concentration of absorbing species in the sample. These techniques are known as cavity ring down (CRD) or cavity enhanced absorption (CEA). Similar to linear absorption techniques the relative measurement uncertainties are often $> 10^{-4}$, i.e. k_{min} is not much less than 10^{-10} cm^{-1} [159-162]. Another order of magnitude in sensitivity may be gained by frequency locking of the laser to the cavity [163]. Comprehensive reviews including recent developments and extensions can be found in [163-165]. An extensive list of measured species, though already published in 2000, is provided in a review of Berden et al. [166].

Additional substantial sensitivity improvement is achieved by combining the basic ideas of FM techniques with the advantage of increased absorption lengths in optical resonators ($k_{min} \sim 10^{-12}$ cm^{-1}) [167]. These modifications to the conventional cavity enhanced absorption method and additional frequency locking of the resonator are known as noise-immune cavity enhanced optical heterodyne molecular spectroscopy (NICE-OHMS) providing the best sensitivity (10^{-14} cm^{-1} or 10^{-9} fractional absorption) reported to date [168]. Specifically this technique has recently been reviewed [169]. Present developments aimed at resonantly coupling of optical frequency combs with cavities resulting in highly sensitive high resolution measurements of broad spectral coverage [170,171 and references therein].

Spectroscopic Technique	$(\Delta I/I_0)_{min}$	k_{min} [cm^{-1}]	n_{min} [cm^{-3}]
Single-Pass (direct)	$10^{-2} \dots 10^{-4}$	$10^{-3} \dots 10^{-5}$	10^{14}
Single-Pass (WM, averaging, FM)	$10^{-5} \dots 10^{-7}$	$10^{-6} \dots 10^{-8}$	$10^{13} \dots 10^{12}$
Multi-Pass (direct)	$10^{-2} \dots 10^{-4}$	$10^{-6} \dots 10^{-8}$	10^{12}
Multi-Pass (WM, averaging, FM)	$10^{-5} \dots 10^{-7}$	$10^{-9} \dots 10^{-11}$	$10^{11} \dots 10^{9}$
CRD (pulsed)	$10^{-2} \dots 10^{-3}$	$10^{-6} \dots 10^{-10}$	10^{10}
CRD/CEA (cw, locked)	$10^{-3} \dots 10^{-5}$	$10^{-8} \dots 10^{-12}$	10^{9}
NICE-OHMS	$> 10^{-8}$	$10^{-11} \dots 10^{-14}$	$10^{9} \dots 10^{6}$

Table 2.2: Sensitivity comparison of different absorption techniques. Acronyms are defined in the text. The MDND (n_{min}) was estimated for a typical line strength of MIR fundamental molecular transitions ($S = 10^{-20}$ cm/molecule) and 1 s integration times. Thus, especially the MDND for methods based on optical resonators may be considered as prediction since measurements have mainly been performed in the weaker overtone region so far.

It is clear from table 2.2 that single pass methods are sufficient for detecting precursors and main products in reactive low pressure plasmas whereas multi-pass arrangements are required for by-products, transient species and main radicals [172,173]. The individual MDND strongly depends on the selected transition and external parameters, among them gas temperature and pressure. The detection limits of several molecular species in the near and mid infrared spectral range achieved by using multi-pass cells in combination with tuneable semiconductor lasers at elevated pressures are listed in [2,174-176]. Taking into account gas temperatures in the plasma being above room temperature, which often results in reduced line strengths, the application of optical resonators should be considered for measuring important radicals. Such measurements have mainly been performed on electronic and overtone transitions so far [177-180 and references therein] which was also a consequence of the available light sources. The application of optical cavities in the molecular fingerprint region could thus decrease the detection limit in *in-situ* plasma diagnostics further.

A   Appendix

A.1   Spectral coverage of selected infrared light sources

Type	Spectral coverage [µm]			Output Power [a]	Reference
Solid State & Colour Centre Lasers					
(Doped) Solid State Lasers					
Tm^{3+}	1.8	...	2.5	10 W	[8]
Ho^{3+}	1.9	...	2.1	1 W	[8]
Ho^{3+}	2.8	...	3.1	1 W	[8]
Er^{3+}	2.6	...	2.9	1 W	[8]
Cr^{2+}:ZnSe (ZnS)	1.5	...	2.4 (3.1)	1 W	[8]
Fe^{2+}:ZnSe	3.9	...	5.0	1 W	[8]
Colour Centre Lasers					
NaCl	1.4	...	1.8	1 W	[6]
KCl:Tl	1.4	...	1.6	1 W	[6]
KCl:Na	1.6	...	1.9	10 mW	[6]
	2.2	...	2.6	35 mW	[6]
KCl:Li	2.0	...	2.5	25 mW	[6]
	2.5	...	2.9	240 mW	[6]
RbCl:Li	2.6	...	3.3	50 mW	[6]
RbJ:Li	2.6	...	3.8		[6]
Optical Parametric Sources [b]					
PPLN, PPKT, PPLT	2	...	5		[8]
$ZnGeP_2$ (ZGP)	4	...	8		[8]
$AgGaS_2$ (AGS)	2	...	9		[8]
$AgGaSe_2$ (AGSe)	3	...	12		[8]
QPM GaAs	2.5	...	12		[8]
Diode Lasers (III-V)					
AlGaAs/GaAs	0.6	...	1.1		[6]
InGaAs/InGa(Al,P)As	1.0	...	2.5		[9,94]
InGaAs/GaAsSb	1.7				[9,181]
InGaAsSb/AlGaAsSb	1.9	...	2.7	1 W	[8-10,182]
InGaAsN/GaAs, InP	2	...	5		[9]
InAsSb/InAs	3	...	5	10 mW	[7]
InAsSb/InAsP	3	...	5	10 mW	[8,10]
InAsSb/InAlAsSb	3	...	5	10 mW	[8,10]
InGaAsPSb/InAsPSb	2.7	...	3.8		[183,184]
InGaSb/InAlSb	3	...	4		[185]
InGaSb/AlGaInSb	2	...	3.5		[9]
InGaAsSb/(In)GaSb (type II)	2	...	3.5		[8-10]
InGaSb/InAs (type III)	2.3	...	3.7	10 mW	[7,9,10]
InAlSb/InSb	5				[8,10,186]
Lead Salt Diode Lasers (IV-VI) [c]					
*PbCd*S	2.8	...	4.2	0.3 mW	[6]
Pb*S*S*e*	4.0	...	8.5	0.5 mW	[6]
*PbSn*Te	6.5	...	32	0.2 mW	[6]
*PbSn*Se	8.5	...	32	0.2 mW	[6]
Pb*SnSe*	4	...	8	< 1 mW	[94]
PbSnSeTe	6	...	30	< 1mW	[6]

CdHgTe	3	...	18	< 1 mW	[94]
PbEuSeTe	3.3	...	5.8	0.5 mW	[6]
Cascade Lasers [d]					
Intersubband Lasers (QCL)					
InAlAs/InGaAs (@InP)	> 3.4				[8]
GaAs/AlGaAs (@GaAs)	> 8.0				[8,58]
InAs/AlSb (@GaSb, InAs)	3.0	...	3.5		[62,63,187]
InAs/AlAsSb	3.0	...	4.0		[188]
InGaAs/AlAs (@InP)	3.7	...	4.2		[189]
InGaAs/AlAsSb	3.0	...	4.5		[190]
InGaAs/AlGaAsSb	3.6	...	4.9		[191,192]
Si/SiGe	7.0				[193]
ZnCdSe/ZnCdMgSe (@InP)	4.8				[194]
GaN/AlGaN	1.5	...	4.2		[64]
Interband Lasers (ICL)	3.0	...	3.9		[70-73,195]

Table A.1:   Intercomparison of common infrared light sources and their spectral coverage.
[a] output power should be considered as a typical upper level (if available in the literature)
[b] output power may vary from μW to W; depends on mode of operation and signal type (idler, signal)
[c] fractions of the *italic* components are varied, e.g. $Pb_{1-x}Sn_x Te$
[d] in some cases the substrate material is indicated (@...); average output power is usually > 1 mW

A.2   Line profiles

A Lorentz profile being used to describe homogeneously broadened spectral lines can be defined on a wavenumber scale, i.e. ϕ is in cm^{-1}, by

$$\phi_L(\nu - \nu_0) = \frac{1}{\pi} \cdot \frac{\Delta \nu_L}{\Delta \nu_L^2 + (\nu - \nu_0)^2} \,. \qquad (A - 1)$$

The maximum is $1/(\pi \Delta \nu_L)$ and $\Delta \nu_L$ (in cm^{-1}) denotes the HWHM. Inhomogeneous line broadening is described by means of a Gaussian profile. The general form is

$$\phi_G(\nu - \nu_0) = \frac{1}{\chi \sqrt{2\pi}} \exp\left(-\frac{(\nu - \nu_0)^2}{2\chi^2} \right) \qquad (A - 2)$$

where χ is the standard deviation of the Gaussian distribution. Since the standard deviation and the HWHM $\Delta \nu_G$ are linked by $\Delta \nu_G = (2 \cdot \ln 2)^{1/2} \cdot \chi$, the line profile for spectroscopic purposes is better expressed by

$$\phi_G(\nu - \nu_0) = \frac{\sqrt{\ln 2}}{\Delta \nu_G \sqrt{\pi}} \exp\left(-(2 \ln 2) \frac{(\nu - \nu_0)^2}{2\Delta \nu_G^2} \right). \qquad (A - 3)$$

The maximum is $1/\Delta \nu_G \cdot (\ln 2/\pi)^{1/2}$. All profiles are normalised such as that condition (2 - 8) is fulfilled.

A.3   Line strength and transition probabilities

The expressions for line strength and transition probabilities in section 2.2.4 were given in units related to the *HITRAN* (and also *GEISA*) database, i.e. adapted to the cgs system. The corresponding equations in SI notation are listed below.

The Einstein coefficient for induced absorption B_{lu} (cf. eq. (2 - 22)) in SI units is

$$B_{lu} = \frac{2\pi^2}{3\varepsilon_0 h^2} \left\{ \frac{1}{g_l} \mid \mu_{lu} \mid^2 \right\},$$ (A - 4)

and the transition probability for spontaneous emission (cf. eq. (2 - 23)) yields now

$$A_{ul} = 8\pi h \left(\frac{f_{lu}}{c} \right)^3 \cdot \frac{g_l}{g_u} B_{lu} = \frac{16\pi^3}{3\varepsilon_0 h} \left(\frac{f_{lu}}{c} \right)^3 \left\{ \frac{1}{g_u} \mid \mu_{lu} \mid^2 \right\}.$$ (A - 5)

Here f_{lu} is the transition frequency (in Hz) and ε_0 the dielectric constant. Transforming the line strength defined by equation (2 - 17) into the SI system gives

$$S = S_{lu} = \frac{2\pi^2}{3\varepsilon_0 hc} \left(\frac{f_{lu}}{c} \right) \frac{n_l}{n} \cdot \left[1 - \frac{g_l n_u}{g_u n_l} \right] \cdot \left\{ \frac{1}{g_l} \mid \mu_{lu} \mid^2 \right\}.$$ (A - 6)

In thermal equilibrium equation (2 - 18) can be used to express the population densities by the Boltzmann law. The transition dipole moment follows to (cf. eq. (2 - 24))

$$\mid \mu_{lu} \mid^2 = S_{lu} \frac{3\varepsilon_0 hc}{2\pi^2} \cdot \frac{Q(T)}{(f_{lu}/c)} \cdot \frac{1}{\exp(-E_l/k_B T)[1 - \exp(-E_{lu}/k_B T)]}$$ (A - 7)

and leads to a relation linking A_{ul} and the line strength S_{lu} (cf. eq. (2 - 25))

$$A_{ul} = 8\pi \left(\frac{f_{lu}}{c} \right)^2 \cdot c \cdot \frac{Q(T)}{\exp(-E_l/k_B T)[1 - \exp(-E_{lu}/k_B T)]} \cdot \frac{1}{g_u} S_{lu}.$$ (A - 8)

The isotopic abundance was omitted here. Similar equations might be found in [91,92], but note that Demtröder defines the line strength as the squared transition dipole moment. Furthermore the equations (A - 6 ... A - 8) are still provided such as that the absorption coefficient $k(\nu)$ is integrated over a wavenumber scale (eq. (2 - 6)). Consequently, an additional factor $1/c$ appears in S_{lu} and the conversion formulae into μ_{lu} and A_{ul} have to be multiplied by c. In contrast to Demtröder the absorption coefficient and hence S_{lu} already includes the population difference between the lower and the upper level.

Bibliography

[1]   P. Crozet, A.J. Ross, M. Vervloet, *Annu. Rep. Prog. Chem. Sect. C* **98**, 33 (2002).

[2]   J. Röpcke, G. Lombardi, A. Rousseau, P.B. Davies, *Plasma Sources Sci. Technol.* **15**, S148 (2006).

[3]   D. Oepts, A.F.G. van der Meer, P.W. van Amersfoort, *Infrared Phys. Technol.* **36**, 297 (1995).

[4]   G.M.H. Knippels, R.F.X.A.M. Mols, A.F.G. van der Meer, D. Oepts, P.W. van Amersfoort, *Phys. Rev. Lett.* **75**, 1755 (1995).

[5]   F.K. Tittel, D. Richter, A. Fried, *Topics Appl. Phys.* **89**, 445 (2003).

[6]   F.K. Kneubühl, M.W. Sigrist, *Laser*, ISBN 3-519-43032-0, (Teubner, Leipzig, 1999). (*in German*)

[7]   R.F. Curl, F.K. Tittel, *Annu. Rep. Prog. Chem. Sect. C* **98**, 219 (2002).

[8]   A. Godard, *C. R. Physique* **8**, 1100 (2007).

[9]   W. Lei, C. Jagadish, *J. Appl. Phys.* **104**, 091101 (2008).

[10]   A. Joullié, P. Christol, *C. R. Physique* **4**, 621 (2003).

[11]   L.A. Eyres, P.J. Tourreau, T.J. Pinguet, C.B. Ebert, J.S. Harris, M.M. Fejer, L. Becouarn, B. Gerard, E. Lallier, *Appl. Phys. Lett.* **79**, 904 (2001).

[12]   S.E. Bisson, T.J. Kulp, O. Levi, J.S. Harris, M.M. Fejer, *Appl. Phys. B* **85**, 199 (2006).

[13]   L. Esaki, *IEEE J. Quantum Electron.* **22**, 1611 (1986).

[14]   M. Osinski, J. Buus, *J. Quantum Electron.* **23**, 9 (1987).

[15]   A.W. Mantz, *Spectrochim. Acta Part A* **51**, 2211 (1995).

[16]   P. Werle, K. Maurer, R. Kormann, R. Mücke, F. D'Amato, T. Lancia, A. Popov, *Spectrochim. Acta Part A* **58**, 2361 (2002).

[17]   Z. Shi, M. Tacke, A. Lambrecht, H. Böttner, *Appl. Phys. Lett.* **66**, 2537 (1995).

[18]   M. Eibelhuber, T. Schwarzl, G. Springholz, W. Heiss, *Appl. Phys. Lett.* **94**, 021118 (2009).

[19]   Z. Feit, D. Kostyk, R.J. Woods, P. Mak, *Appl. Phys. Lett.* **58**, 343 (1991).

[20]   M. Tacke, *Infrared Phys. Technol.* **36**, 447 (1995).

[21]   P.J. McCann, Y. Selivanov, *Mater. Res. Soc. Symp. Proc.* **891**, 0891-EE01-05 (2006).

[22]   J. Faist, C. Sirtori, F. Capasso, L. Pfeiffer, K.W. West, *Appl. Phys. Lett.* **64**, 872 (1994).

[23]   L. Esaki, R. Tsu, *IBM J. Res. Develop.* **14**, 61 (1970).

[24]   R.F. Kazarinov, R.A. Suris, *Sov. Phys. Semicond.* **5**, 707 (1971).

[25]   F. Capasso, K. Mohammed, A.Y. Cho, *IEEE J. Quantum Electron.* **22**, 1853 (1986).

[26]   P. Yuh, K.L. Wang, *Appl. Phys. Lett.* **51**, 1404 (1987).

[27]   M. Helm, P. England, E. Colas, F. DeRosa, S.J. Allen, *Phys. Rev. Lett.* **63**, 74 (1989).

[28]   J. Faist, F. Capasso, C. Sirtori, D. sivco, A.K. Hutvhinson, S.G. Chu, A.Y. Cho, *Appl. Phys. Lett.* **64**, 1144 (1994).

[29]   J. Faist, F. Capasso, D.L. Sivco, C. Sirtori, A.L. Hutvhinson, A.Y. Cho, *Science* **264**, 553 (1994).

[30]   R. Tsu, *Nature* **369**, 442 (1994).

[31]   Y. Bai, S.R. Darvish, S. Slivken, P. Sung, J. Nguyen, A. Evans, W. Zhang, M. Razeghi, *Appl. Phys. Lett.* **91**, 141123 (2007).
[32]   Y. Bai. B. Gokden, S. Slivken, S.R. Darvish, S.A. Pour, M. Razeghi, *Proc. SPIE* **7222**, 72220O (2009).
[33]   J. Faist, F. Capasso, C. Sirtori, D.L. Sivco, A.L. Hutchinson, M.S. Hybertsen, A.Y. Cho, *Phys. Rev. Lett.* **76**, 411 (1996).
[34]   C. Sirtori, J. Faist, F. Capasso, D.L. Sivco, A.L. Hutchinson, A.Y. Cho, *IEEE J. Quantum Electron.* **33**, 89 (1997).
[35]   D. Hofstetter, M. Beck, T. Aellen, J. Faist, *Appl. Phys. Lett.* **78**, 396 (2001).
[36]   Q.J. Wang, C. Pflügl, L. Diehl, F. Capasso, T. Edamura, S. Furuta, M. Yamanishi, H. Kan, *Appl. Phys. Lett.* **94**, 011103 (2009).
[37]   J. Faist, M. Beck, T. Aellen, E. Gini, *Appl. Phys. Lett.* **78**, 147 (2001).
[38]   G. Scamarcio, C. Gmachl, F. Capasso, A. Tredicucci, A.L. Hutchinson, D.L. Sivco, A.Y Cho, *Semicond. Sci. Technol.* **13**, 1333 (1998).
[39]   A. Tredicucci, F. Capasso, C. Gmachl, D.L. Sivco, A.L. Hutchinson, A.Y. Cho, *Appl. Phys. Lett.* **73**, 2101 (1998).
[40]   A. Tredicucci, F. Capasso, C. Gmachl, D.L. Sivco, A.L. Hutchinson, A.Y. Cho, J. Faist, G. Scamarcio, *Appl. Phys. Lett.* **72**, 2388 (1998).
[41]   F. Capasso, A. Tredicucci, C. Gmachl, D.L. Sivco, A.L. Hutchinson, A.Y. Cho, G. Scamarcio, *IEEE J. Sel. Top. Quantum Electron.* **5**, 792 (1999).
[42]   J. Faist, F. Capasso, C. Sirtori, D.L. Sivco, A.L. Hutchinson, A.Y. Cho, *Nature* **387**, 777 (1997).
[43]   R.Q. Yang, *Supperlat. Microstruct.* **17**, 77 (1995).
[44]   M.C. Wanke, F. Capasso, C. Gmachl, A. Tredicucci, D.L. Sivco, A.L. Hutchinson, S.G. Chu, A.Y. Cho, *Appl. Phys. Lett.* **78**, 3950 (2001).
[45]   C. Gmachl, F. Capasso, A. Tredicucci, D.L. Sivco, A.L. Hutchinson, S.G. Chu, A.Y. Cho, *Appl. Phys. Lett.* **73**, 3830 (1998).
[46]   F. Capasso, C. Gmachl, R. Paiella, A. Tredicucci, A.L. Hutchinson, D.L. Sivco, J.N. Baillargeon, A.Y. Cho, H.C. Liu, *IEEE J. Sel. Top. Quantum Electron.* **6**, 931 (2000).
[47]   J. Faist, F. Capasso, C. Sirtori, D.L. Sivco, J.N. Baillargeon, A.L. Hutchinson, S.G. Chu, A.Y. Cho, *Appl. Phys. Lett.* **68**, 3680 (1996).
[48]   D. Hofstetter, M. Beck, T. Aellen, J. Faist, U. Oesterle, M. Ilegems, E. Gini, H. Melchior, *Appl. Phys. Lett.* **78**, 1964 (2001).
[49]   A. Tredicucci, C. Gmachl, F. Capasso, D.L. Sivco, A.L. Hutchinson, A.Y. Cho, *Appl. Phys. Lett.* **74**, 638 (1999).
[50]   C. Gmachl, F. Capasso, D.L. Sivco, A.Y. Cho, *Rep. Prog. Phys.* **64**, 1533 (2001).
[51]   J. Faist, D. Hofstetter, M. Beck, T. Aellen, M. Rochat, S. Blaser, *IEEE J. Quantum Electron.* **38**, 533 (2002).
[52]   F. Capasso, C. Gmachl, D.L. Sivco, A.Y. Cho, *Phys. Today* **55**, 34 (2002).
[53]   C. Sirtori, J. Nagle, *C. R. Physique* **4**, 639 (2003).
[54]   A. Evans, S.R. Darvish, S. Slivken, J. Nguyen, Y. Bai, M. Razeghi, *Appl. Phys. Lett.* **91**, 071101 (2007).
[55]   A. Bismuto, T. Gresch, A. Bächle, J. Faist, *Appl. Phys. Lett.* **93**, 231104 (2008).
[56]   Y. Bai, S.R. Darvish, S. Slivken, W. Zhang, A. Evans, J. Nguyen, M. Razeghi, *Appl. Phys. Lett.* **92**, 101105 (2008).
[57]   Y. Bai, S. Slivken, S.R. Darvish, M. Razeghi, *Appl. Phys. Lett.* **93**, 021103 (2008).

[58]   C. Sirtori, P. Kruck, S. Barbieri, P. Collot, J. Nagle, M. Beck, J. Faist, U. Oesterle, *Appl. Phys. Lett.* **73**, 3486 (1998).

[59]   C. Sirtori, H. Page, C. Becker, V. Ortiz, *IEEE J. Quantum Electron.* **38**, 547 (2002).

[60]   J. Faist, F. Capasso, D.L. Sivco, A.L. Hutchinson, S.G. Chu, A.Y. Cho, *Appl. Phys. Lett.* **72**, 680 (1998).

[61]   I. Vurgaftman, J.R. Meyer, L.R. Ram-Mohan, *J. Appl. Phys.* **89**, 5851 (2001).

[62]   R. Teissier, D. Barate, A. Vicet, C. Alibert, A. N. Baranov, X. Marcadet, C. Renard, M. Garcia, C. Sirtori, D. Revin, J. Cockburn, *Appl. Phys. Lett.* **85**, 167 (2004).

[63]   J. Devenson, O. Cathabard, R. Teissier, A. N. Baranov, *Appl. Phys. Lett.* **91**, 251102 (2007).

[64]   A.Y. Cho, D.L. Sivco, H.M. Ng, C. Gmachl, A. Tredicucci, A.L. Hutchinson, S.G. Chu, F. Capasso, *J. Cryst. Growth* **227-228**, 1 (2001).

[65]   J.R. Meyer, I. Vurgaftman, R.Q. Yang, L.R. Ram-Mohan, *IEEE Electron. Lett.* **32**, 45 (1996).

[66]   I. Vurgaftman, J.R. Meyer, L.R. Ram-Mohan, *IEEE Phot. Technol. Lett.* **9**, 170 (1997).

[67]   J.R. Meyer, C.A. Hoffman, F.J. Bartoli, L.R. Ram-Mohan, *Appl. Phys. Lett.* **67**, 757 (1995).

[68]   R.Q. Yang, S.S. Pei, *J. Appl. Phys.* **79**, 8197 (1996).

[69]   L.J. Olafsen, E.H. Aifer, I. Vurgaftman, W.W. Bewley, C.L. Felix, J.R. Meyer, D. Zhang, C.H. Lin, S.S. Pei, *Appl. Phys. Lett.* **72**, 2370 (1998).

[70]   C.L. Felix, W.W. Bewley, E.H. Aifer, I. Vurgaftman, J.R. Meyer, C.H. Lin, D. Zhang, S.J. Murry, R.Q. Yang, S.S. Pei, *J. Electron. Mater.* **27**, 77 (1998).

[71]   C.L. Felix, W.W. Bewley, I. Vurgaftman, J.R. Meyer, D. Zhang, C.H. Lin, R.Q. Yang, S.S. Pei, *IEEE Phot. Technol. Lett.* **9**, 1433 (1997).

[72]   C.H. Lin, R.Q. Yang, D. Zhang, S.J. Murry, S.S. Pei, A.A. Allerman, S.R. Kurtz, *IEEE Electron. Lett.* **33**, 598 (1997).

[73]   R.Q. Yang, B.H. Yang, D. Zhang, C.H. Lin, S.J. Murry, H. Wu, S.S. Pei, *Appl. Phys. Lett.* **71**, 2409 (1997).

[74]   R.Q. Yang, J.L. Bradshaw, J.D. Bruno, J.T. Pham, D.E. Wortman, *IEEE J. Quantum Electron.* **38**, 559 (2002).

[75]   G. Totschnig, F. Winter, V. Pustogov, J. Faist, A. Müller, *Opt. Lett.* **27**, 1788 (2002).

[76]   G.P. Luo, C. Peng, H.Q. Le, S.S. Pei, W.Y. Hwang, B. Ishaug, J. Um, J.N. Baillargeon, C.H. Lin, *Appl. Phys. Lett.* **78**, 2834 (2001).

[77]   G. Luo, C. Peng, H.Q. Le, S.S. Pei, H. Lee, W.Y. Hwang, B. Ishaug, J. Zheng, *IEEE J. Quantum Electron.* **38**, 486 (2002).

[78]   R. Maulini, M. Beck, J. Faist, E. Gini, *Appl. Phys. Lett.* **84**, 1659 (2004).

[79]   R. Maulini, A. Mohan, M. Giovannini, J. Faist, E. Gini, *Appl. Phys. Lett.* **88**, 201113 (2006).

[80]   G. Wysocki, R. Lewicki, R.F. Curl, F.K. Tittel, L. Diehl, F. Capasso, M. Troccoli, G. Hofler, D. Bour, S. Corzine, R. Maulini, M. Giovannini, J. Faist, *Appl. Phys. B* **92**, 305 (2008).

[81]   J. Faist, C. Gmachl, F. Capasso, C. Sirtori, D.L. Sivco, J.N. Baillargeon, A.Y. Cho, *Appl. Phys. Lett.* **70**, 2670 (1997).

[82]   J. Koeth, M. Fischer, M. Legge, J. Seufert, R. Werner, *Photonik*, 36 (1/2005). (*in German*).

[83]   C. Gmachl, J. Faist, J.N. Baillargeon, F. Capasso, C. Sirtori, D.L. Sivco, S.G. Chu, A.Y. Cho, *IEEE Phot. Technol. Lett.* **9**, 1090 (1997).

[84]   E. Normand, G. Duxbury, N. Langford, *Opt. Comm.* **197**, 115 (2001).

[85]   F. Capasso, J. Faist, C. Sirtori, A.Y. Cho, *Solid State Comm.* **102**, 231 (1997).

[86]   S.S. Penner, *Quantitative Molecular Spectroscopy and Gas Emissivities*, ISBN 978-020105760-7, (Addison-Wesley, Reading, 1959).

[87]   G. Herzberg, *Molecular Spectra and Molecular Structure. I. Spectra of Diatomic Molecules*, ISBN 0-89464-270-7, (Krieger Publishing, Malabar, 1989).

[88]   M. Simeckova, D. Jacquemart, L.S. Rothman, R.R. Gamache, A. Goldman, *J. Quant. Spectrosc. Radiat. Transfer* **98**, 130 (2006).

[89]   A. Goldman, R.R. Gamache, A. Perrin, J.M. Flaud, C.P. Rinsland, L.S. Rothman, *J. Quant. Spectrosc. Radiat. Transfer* **66**, 455 (2000).

[90]   L.S. Rothman, C. P. Rinsland, A. Goldman, S.T. Massie, D.P. Edwards, J.M. Flaud, A. Perrin, C. Camy-Peyret, V. Dana, J.Y. Mandin, J. Schroeder, A. McCann, R.R. Gamache, R.B. Wattson, K. Yoshino, K.V. Chance, K.W. Jucks, L.R. Brown, V. Nemtchinov, P. Varanasi, *J. Quant. Spectrosc. Radiat. Transfer* **60**, 665 (1998).

[91]   W. Demtröder, *Laserspektroskopie*, ISBN 3-540-64219-6, (Springer, Berlin, 2004). (*in German*)

[92]   W. Demtröder, *Laser Spectroscopy*, ISBN 3-540-65225-6, (Springer, Berlin, 2003).

[93]   W.S. Letochow, *Laserspektroskopie*, ISBN 978-352806830-1, (Akademie Verlag, Berlin, 1977). (*in German*)

[94]   G. Duxbury, *Infrared Vibration-Rotation Spectroscopy. From Free Radicals to the Infrared Sky*, ISBN 0-471-97419-6, (John Wiley & Sons, Chichester, 2000).

[95]   P.M. Chu, F.R. Guenther, G.C. Rhoderick, W.J. Lafferty, *J. Res. Natl. Inst. Stand. Technol.* **104**, 59 (1999).

[96]   S.W. Sharpe, T.J. Johnson, R.L. Sams, P.M. Chu, G.C. Rhoderick, P.A. Johnson, *Appl. Spectrosc.* **58**, 1452 (2004).

[97]   G.D. Stancu, J. Röpcke, P.B. Davies, *Journ. Chem. Phys.* **122**, 014306 (2005).

[98]   L.S. Rothman, N. Jacquinet-Husson, C. Boulet, A.M. Perrin, *C.R. Physique* **6**, 897 (2005).

[99]   R.A. McClatchey, W.S. Benedict, S.A. Clough, D.E. Burch, R.F. Calfee, K. Fox, L.S. Rothman, J.S. Garing, *Environm. Res. Pap.* **434**, Air Force Cambridge Research Laboratories (AFCRL) Technical Report (TR) 73-0096 (1973).

[100]   N. Husson, B. Bonnet, N.A. Scott, A. Chedin, *J. Quant. Spectrosc. Radiat. Transfer* **48**, 509 (1992).

[101]   L.S. Rothman, A. Goldman, J.R. Gillis, R.R. Gamache, H.M. Pickett, R.L. Poynter, N. Husson, A. Chedin, *Appl. Opt.* **22**, 1616 (1983).

[102]   L.S. Rothman, R.R. Gamache, A. Barbe, A. Goldman, J.R. Gillis, L.R. Brown, R.A. Toth, J.M. Flaud, C. Camy-Peyret, *Appl. Opt.* **22**, 2247 (1983).

[103]   L.S. Rothman, R.R. Gamache, A. Goldman, L.R. Brown, R.A. Toth, H.M. Pickett, R.L. Poynter, J.M. Flaud, C. Camy-Peyret, A. Barbe, N. Husson, C.P. Rinsland, M.A.H. Smith, *Appl. Opt.* **26**, 4058 (1987).

[104]   L.S. Rothman, R.R. Gamache, R.H. Tipping, C.P. Rinsland, M.A.H. Smith, D.C. Benner, V. Malathy Devi, J.M. Flaud, C. Camy-Peyret, A. Perrin, A. Goldman, S.T. Massie, L.R. Brown, R.A. Toth, *J. Quant. Spectrosc. Radiat. Transfer* **48**, 469 (1992).

[105]   L.S. Rothman *et al.*, *J. Quant. Spectrosc. Radiat. Transfer* **82**, 5 (2003).

[106]   L.S. Rothman *et al.*, *J. Quant. Spectrosc. Radiat. Transfer* **96**, 139 (2005).
[107]   N. Husson, B. Bonnet, A. Chedin, N.A. Scott, A.A. Chursin, V.F. Golovko, V.G. Tyuterev, *J. Quant. Spectrosc. Radiat. Transfer* **52**, 425 (1994).
[108]   N. Jacquinet-Husson *et al.*, *J. Quant. Spectrosc. Radiat. Transfer* **62**, 205 (1999).
[109]   N. Jacquinet-Husson *et al.*, *J. Quant. Spectrosc. Radiat. Transfer* **95**, 429 (2005).
[110]   N. Jacquinet-Husson *et al.*, *J. Quant. Spectrosc. Radiat. Transfer* **109**, 1043 (2008).
[111]   H.M. Pickett, R.L. Poynter, E.A. Cohen, M.L. Delitsky, J.C. Pearson, H.S.P. Müller, *J. Quant. Spectrosc. Radiat. Transfer* **60**, 883 (1998).
[112]   D.G. Feist, *J. Quant. Spectrosc. Radiat. Transfer* **85**, 57 (2004).
[113]   K. Chance, K.W. Jucks, D.G. Johnson, W.A. Traub, *J. Quant. Spectrosc. Radiat. Transfer* **52**, 447 (1994).
[114]   H.S.P. Müller, F. Schlöder, J. Stutzki, G. Winnewisser, *J. Molec. Struct.* **742**, 215 (2005).
[115]   H.S.P. Müller, S. Thorwirth, D.A. Roth, G. Winnewisser, *Astron. Astrophys.* **370**, L49 (2001).
[116]   L.R. Brown, C.B. Farmer, C.P. Rinsland, R.A. Toth, *Appl. Opt.* **26**, 5154 (1987).
[117]   R.H. Norton, C.P. Rinsland, *Appl. Opt.* **30**, 389 (1991).
[118]   L.R. Brown, M.R. Gunson, R.A. Toth, F.W. Irion, C.P. Rinsland, A. Goldman, *Appl. Opt.* **35**, 2828 (1996).
[119]   C.B. Farmer, *Mikrochim. Acta* III, 189 (1987).
[120]   R.R. Gamache, L.S. Rothman, *J. Quant. Spectrosc. Radiat. Transfer* **48**, 519 (1992).
[121]   J. Fischer, R.R. Gamache, *J. Quant. Spectrosc. Radiat. Transfer* **74**, 263 (2002).
[122]   J. Fischer, R.R. Gamache, *J. Quant. Spectrosc. Radiat. Transfer* **74**, 273 (2002).
[123]   J. Fischer, R.R. Gamache;, A. Goldman, L.S. Rothman, A. Perrin, *J. Quant. Spectrosc. Radiat. Transfer* **82**, 401 (2003).
[124]   T.G. Schmalz, W.H. Flygare, *Coherent Transient Microwave Spectroscopy and Fourier Transform Methods* in: J.I. Steinfeld, *Laser and Coherence Spectroscopy*, ISBN 0-306-31027-9, (Plenum New York, 1978).
[125]   G. Duxbury, N. Langford, M.T. McCulloch, S. Wright, *Chem. Soc. Rev.* **34**, 921 (2005).
[126]   M.T. McCulloch, G. Duxbury, N. Langford, *Mol. Phys.* **104**, 2767 (2006).
[127]   M.D. Levenson, S.S. Kano, *Introduction to Nonlinear Laser Spectroscopy*, ISBN 0-12-444722-8, (Academic Press, San Diego, 1988).
[128]   F. Bloch, *Phys. Rev.* **70**, 460 (1946).
[129]   R.R. Ernst, *Sensitivity Enhancement in Magnetic Resonance* in: J.S. Waugh *Advances in Magnetic Resonance. Vol. 2*, (Academic Press, New York, 1966).
[130]   B.A. Jacobsohn, R.K. Wangsness, *Phys. Rev.* **73**, 942 (1948).
[131]   N. Bloembergen, E.M. Purcell, R.V. Pound, *Phys. Rev.* **73**, 679 (1948).
[132]   A.M. Portis, *Phys. Rev.* **100**, 1219 (1955).
[133]   J.W. Stoner, D. Szymanski, S.S. Eaton, R.W. Quine, G.A. Rinard, G.R. Eaton, *J. Magn. Reson.* **170**, 127 (2004).
[134]   S. Zamith, J. Degert, S. Stock, B. de Beauvoir, V. Blanchet, M.A. Bouchene, B. Girard, *Phys. Rev. Lett.* **87**, 033001 (2001).
[135]   J. S. Melinger, S.R. Gandhi, A. Hariharan, D. Goswami, W.S. Warren, *J. Chem. Phys.* **101**, 6439 (1994).
[136]   M.M.T. Loy, *Phys. Rev. Lett.* **32**, 814 (1974).

[137]   S.M. Hamadani, A.T. Mattick, N.A. Kurnit, A. Javan, *Appl. Phys. Lett.* **27**, 21 (1975).
[138]   J.C. McGurk, T.G. Schmalz, W.H. Flygare, *J. Chem. Phys.* **60**, 4181 (1974).
[139]   J.C. McGurk, H. Mäder, R.T. Hofmann, T.G. Schmalz, W.H. Flygare, *J. Chem. Phys.* **61**, 3759 (1974).
[140]   B. Macke, P. Glorieux, *Chem. Phys. Lett.* **14**, 85 (1972).
[141]   V.V. Khodos, D.A. Ryndyk, V.L. Vaks, *Eur. Phys. J. Appl. Phys.* **25**, 203 (2004).
[142]   C. Liedenbaum, S. Stolte, J. Reuss, *Phys. Rep.* **178**, 1 (1989).
[143]   P. Werle, F. Slemr, M. Gehrtz, C. Bräuchle, *Appl. Phys. B* **49**, 99 (1989).
[144]   G. Friedrichs, *Z. Phys. Chem.* **222**, 1 (2008).
[145]   J. Reid, D. Labrie, *Appl. Phys. B* **26**, 203 (1981).
[146]   D.S. Bomse, A.C. Stanton, J.A. Silver, *Appl. Opt.* **31**, 718 (1992).
[147]   M.S. Zahniser, D.D. Nelson, J.B. McManus, P.L. Kebabian, *Phil. Trans. R. Soc. Lond. A* **351**, 371 (1995).
[148]   C.V. Horii, M.S. Zahniser, D.D. Nelson, J.B. McManus, S.C. Wofsy, *Proc. SPIE* **3758**, 152 (1999).
[149]   P. Werle, R. Mücke, F. Slemr, *Appl. Phys. B* **57**, 131 (1993).
[150]   H. Linnartz, *Phys. Script.* **70**, C24 (2004).
[151]   J.U. White, *J. Opt. Soc. Am.* **32**, 285 (1942).
[152]   S.M. Chernin, E.G. Barskaya, *Appl. Opt.* **30**, 51 (1991).
[153]   D. Herriott, H. Kogelnik, R. Kompfner, *Appl. Opt.* **3**, 523 (1964).
[154]   D.R. Herriott, H.J. Schulte, *Appl. Opt.* **4**, 883 (1965).
[155]   J.B. McManus, P.L. Kebabian, M.S. Zahniser, *Appl. Opt.* **34**, 3336 (1995).
[156]   A. O'Keefe, D.A.G. Deacon, *Rev. Sci. Instrum.* **59**, 2544 (1988).
[157]   R. Engeln, G. Berden, R. Peeters, G. Meijer, *Rev. Sci. Instrum.* **69**, 3763 (1998).
[158]   A. O'Keefe, *Chem. Phys. Lett.* **293**, 331 (1998).
[159]   R.D. van Zee, J.T. Hodges, J.P. Looney, *Appl. Opt.* **38**, 3951 (1999).
[160]   D. Romanini, A. A. Kachanov, F. Stoeckel, *Chem. Phys. Lett.* **270**, 538 (1997).
[161]   M.D. Levenson, B.A. Paldus, T.G. Spence, C.C. Harb, J.S. Harris, R.N. Zare, *Chem. Phys. Lett.* **290**, 335 (1998).
[162]   E.J. Moyer, D.S. Sayres, G.S. Engel, J.M.S. Clair, F.N. Keutsch, N.T. Allen, J.H. Kroll, J.G. Anderson, *Appl. Phys. B* **92**, 467 (2008).
[163]   B.A. Paldus, A.A. Kachanov, *Can. J. Phys.* **83**, 975 (2005).
[164]   M. Mazurenka, A.J. Orr-Ewing, R. Peverall, G.A.D. Ritchie, *Annu. Rep. Prog. Chem. Sect. C.* **101**, 100 (2005).
[165]   G. Friedrichs, *Z. Phys. Chem.* **222**, 31 (2008).
[166]   G Berden, R. Peeters, G. Meijer, *Int. Rev. Phys. Chem.* **19**, 565 (2000).
[167]   J. Ye, L.S. Ma, J.L. Hall, *Opt. Lett.* **21**, 1000 (1996).
[168]   J. Ye, L.S. Ma, J.L. Hall, *J. Opt. Soc. Am. B* **15**, 6 (1998).
[169]   A. Foltynowicz, F.M. Schmidt, W. Ma, O. Axner, *Appl. Phys. B* **92**, 313 (2008).
[170]   M.J. Thorpe, J. Ye, *Appl. Phys. B* **91**, 397 (2008).
[171]   M.J. Thorpe, F. Adler, K.C. Cossel, M.H.G. de Miranda, J. Ye, *Chem. Phys. Lett.* **468**, 1 (2009).
[172]   F. Hempel, P.B. Davies, D. Loffhagen, L. Mechold, J. Röpcke, *Plasma Sources Sci. Technol.* **12**, S98 (2003).
[173]   L. Mechold, J. Röpcke, X. Duten, A. Rousseau, *Plasma Sources Sci. Technol.* **10**, 52 (2001).

[174]   P. Werle, *Spectrochim. Acta Part A* **54**, 197 (1998).

[175]   M.S. Zahniser, D.D. Nelson, J.B. McManus, S.C. Herndon, E.C. Wood, J.H. Shorter, B.H. Lee, G.W. Santoni, R. Jimenez, B.C. Daube, S. Park, E.A. Kort, S.C.Wofsy, *Proc. SPIE* **7222**, 72220H (2009).

[176]   M.S. Zahniser, presented at the 2nd IPS, Greifswald, Germany, July 2007, Book of Tutorials.

[177]   P. Vankan, T. Rutten, S. Mazouffre, D.C. Schram, R. Engeln, *Appl. Phys. Lett.* **81**, 418 (2002).

[178]   J. Ma, J.C. Richley, M.N.R. Ashfold,Y.A. Mankelevich, *J. Appl. Phys.* **104**, 103305 (2008).

[179]   R. Peeters, G. Berden, G.Meijer, *Appl. Phys. B* **73**, 65 (2001).

[180]   S. Cheskis, *Prog. Energy Comb. Sci.* **25**, 233 (1999).

[181]   J.Y.T Huang, L.J. Mawst, T.F. Kuech, X. Song, S E. Babcock, C.S. Kim, I. Vurgaftman, J.R. Meyer, A.L. Holmes, *J. Phys. D: Appl. Phys.* **42**, 025108 (2009).

[182]   M. Grau, C. Lin, O. Dier, M.C. Amann, *Phys. E* **20**, 507 (2004).

[183]   M. Yin, A. Krier, P.J. Carrington, N. Cook, presented at the 9th Int. Conf. on Mid-Infrared Optoelectronics: Materials and Devices (MIOMD), Freiburg, Germany, 7 - 11 September 2008.

[184]   N. Cook, A. Krier, P. Batty, R. Jones, presented at the 9th Int. Conf. on Mid-Infrared Optoelectronics: Materials and Devices (MIOMD), Freiburg, Germany, 7 - 11 September 2008.

[185]   M. Yin, G.R. Nash, S.D. Coomber, L. Buckle, P.J. Carrington, A. Krier, A. Andreev, S.J.B. Przeslak, G. de Valicourt, S.J. Smith, M.T. Emeny, T. Ashley, *Appl. Phys. Lett.* **93**, 121106 (2008).

[186]   T. Ashley, C.T. Elliott, R. Jefferies, A.D. Johnson, G.J. Pryce, A.M. White, M. Carroll, *Appl. Phys. Lett.* **70**, 931 (1997).

[187]   K. Ohtani, H. Ohno, *Appl. Phys. Lett.* **82**, 1003 (2003).

[188]   X. Marcadet, C. Renard, M. Carras, M. Garcia, J. Massies, *Appl. Phys. Lett.* **91**, 161104 (2007).

[189]   M.P. Semtsiv, M. Ziegler, S. Dressler, W.T. Masselink, N. Georgiev, T. Dekorsy, M. Helm, *Appl. Phys. Lett.* **85**, 1478 (2004).

[190]   S.Y. Zhang, D.G. Revin, J.W. Cockburn, M. Steer, K. Kennedy, A.B. Krysa, M. Hopkinson, presented at the 9th Int. Conf. on Mid-Infrared Optoelectronics: Materials and Devices (MIOMD), Freiburg, Germany, 7 - 11 September 2008.

[191]   Q. Yang, C. Manz, W. Bronner, L. Kirste, K. Köhler, J. Wagner, *Appl. Phys. Lett.* **86**, 131109 (2005).

[192]   Q. Yang, C. Manz, W. Bronner, N. Lehmann, F. Fuchs, K. Köhler, J. Wagner, *Appl. Phys. Lett.* **90**, 121134 (2007).

[193]   L. Diehl, S. Mentes, E. Müller, D. Grützmacher, H. Sigg, U. Gennser, I. Sagnes, Y. Campidelli, O. Kermarrec, D. Bensahel, J. Faist, *Appl. Phys. Lett.* **81**, 4700 (2002).

[194]   K.J. Franz, W.O. Charles, A. Shen, A.J. Hoffman, M.C. Tamargo, C. Gmachl, presented at the 9th Int. Conf. on Mid-Infrared Optoelectronics: Materials and Devices (MIOMD), Freiburg, Germany, 7 - 11 September 2008.

[195]   W.W. Bewley, J.A. Nolde, D.C. Larrabee, C.L. Canedy, C.S. Kim, M. Kim, I. Vurgaftman, J.R. Meyer, *Appl. Phys. Lett.* **89**, 161106 (2006).

3 Molecule conversion in reactive plasmas containing H_2-N_2-O_2

3.1 Motivation

The dissociating and ionising properties of plasmas lead to the generation of a wide variety of species, such as radicals and ions, which are in turn used to form new products via gas phase and plasma surface interactions. Reactive plasmas provide therefore an efficient means for molecule conversion and are used for a variety of applications. A detailed understanding of plasma chemical systems is highly desirable in order to improve the efficiency of the processes and to facilitate the development of new applications.

Apart from fluorocarbon, silane or other technologically relevant plasmas, hydrocarbon containing discharges have been used and studied extensively. Usually thin film deposition in general and diamond deposition in particular, CH_4 reforming and, more recently, dust formation were in the centre of interest. Pure [1-6] or relatively basic gas mixtures of CH_4/H_2 [7-17], CH_4/O_2 [18,19] or $CH_4/H_2/O_2$ [20-22] have been investigated experimentally and theoretically. Other oxygen containing precursors, e.g. methanol [23] have been used as well to study the C - H - O system [24]. Very similar, CH_4/N_2 or $CH_4/H_2/N_2$ mixtures are of increasing interest for thin film deposition, chemical synthesis or astrophysics and are hence investigated [25-28]. Recently, hydrocarbon containing plasmas have attracted even more attention in environmental science and in the fusion community. Specifically in the latter field, the similarity between non-thermal hydrocarbon plasmas and the peripheral regions of fusion devices (divertor) has been recognised which may help tackling the hydrogen isotope co-deposition in such experiments [4,29]. The chemistry in detoxification and plasma assisted volatile organic compound removal approaches as well as in biomedical applications is strongly connected to the C - H - N - O chemical system [30-37]. The relevance is now especially recognised in the latter field employing usually atmospheric pressure plasmas.

The majority of all these non-thermal plasmas were based on direct current (DC) [17], radio frequency (RF) [2,19] or microwave (MW) [13,25] discharges. More recently, microplasmas or plasma jets have been employed [3,36]. Plasma diagnostics usually concentrated on optical emission spectroscopy (OES) [1,9,21,26], gas chromatography (GC) [6,31,34] and mass spectrometry (MS) [3,18,27] in order to obtain information on densities of molecular species and their kinetics. Specifically in CH_4/N_2 gas mixtures the selectivity of MS is rather limited [38] and experimental data require a thorough calibration and post-measurement treatment. In contrast, infrared (tuneable diode) laser absorption spectroscopy (IRLAS) provides a valuable, highly selective means of determining number densities of transient and stable molecular species in their ground state, which are especially relevant for chemical processes [11,14,15,18,22].

Following earlier IRLAS studies of $Ar/CH_4/H_2/O_2$ and $Ar/CH_4/H_2/N_2$ microwave discharges by Mechold and Hempel [39,40], which were accompanied by chemical modelling, this chapter concerns an $Ar/CH_4/N_2/O_2$ mixture. Such chemical systems were also of interest for the NO_x removal from flue gases by selective catalytic or non-catalytic reduction [41]. Due to the CH_4 decomposition this model system should effectively be considered as an $Ar/CH_4/H_2/N_2/O_2$ mixture. It thus represents a combination of both earlier works and is mainly motivated by two aspects:

i)   recent progress in IR-TDLAS instrumentation, and

ii)   forthcoming evidence of the non-negligible role of surface processes for the molecule conversion and gas phase composition in reactive plasmas.

Firstly, as suggested by Mechold [39], the development of the rapid laser scan technology [42] in combination with *in-situ* multiple pass cell optics now enables time resolved experiments and a higher measurement sensitivity to be achieved. According to the model predictions more radicals or intermediate species should be accessible. The focus here is on the hydroxyl radical (OH) being considered as the main oxidising radical, e.g. in combustion processes or atmospheric chemistry. However, OH is often detected in its electronically excited state by OES in the UV [35,36], but less frequently in its ground state, e.g., by laser induced fluorescence [43]. The short lifetime requires a high sensitivity which was commonly achieved by absorption spectroscopy employing optical resonators in the UV [43] and NIR [44,45] respectively. The detection of the hydroxyl radical in its ro-vibrational fundamental band in the MIR using either an FT-IR spectrometer [37] or IRLAS in combination with multiple pass optics has, as far as we are aware, not yet been reported.

Secondly, many of the reported (plasma) chemical models, particularly those being complimentary to the IRLAS studies or to the microwave reactor used for the present experiments, were based on gas phase reactions [14,22,25,28,40] and have demonstrated reasonable agreement with the measurements. Recently, the importance of surface reactions for the abundance of stable products in the gas phase has been pointed out [46-49]. The influence of the surface (material) on the formation of e.g. ammonia was already mentioned in the 1970s [50]. The precursors (N_2, H_2, O_2) used in several such investigations are also relevant for the intended measurements on $Ar/CH_4/(H_2)/N_2/O_2$. The plasma catalytic synthesis of NH_3 from N_2/H_2 discharges has been extensively studied, both experimentally [48,51-53] and theoretically [54,55]. N_2/O_2 plasmas have been examined similarly using both approaches [49,56,57]. The application of sophisticated dosing experiments with either N_2, O_2 or H_2 in the post-discharge revealed the importance of the plasma produced surface state (i.e. coverage with radicals) for the formation of stable products such as NO_2, N_2O and NH_3 [49].

The majority of the above mentioned measurements on the H_2 - N_2 - O_2 chemical system were carried out in an expanding thermal plasma (ETP) in Eindhoven, being in fact a remote plasma, i.e. the energy transfer to the plasma is separated from the gas phase and surface chemistry. The plasma is created in a cascaded arc at elevated pressures ($\sim$ 400 mbar) and expands into a low pressure (typically below 1 mbar) reactor. The feed gas (mixture), e.g. Ar or mixtures of Ar/H_2, Ar/N_2 or N_2/H_2, is efficiently ionised up to 20 % and molecular components are almost fully dissociated. Due to the low electron temperature in the expansion electron impact ionisation or dissociation can be neglected. Decomposition of molecular precursors, that are injected in the background, is governed by charge exchange reactions (mainly with Ar^+) and subsequent dissociative recombination [53,58-60]. The absence of electron induced ionisation and dissociation in the recombining part of the plasma is a key difference to the microwave discharge used here. Modelling of the latter system requires usually much more reactions to be considered. Furthermore, a clear discrimination between gas phase and surface reactions is usually difficult in such plasmas where excitation, ionisation and dissociation are not separated from recombination and surface association as in remote plasmas.

The present approach for the hydrocarbon containing microwave plasma is therefore exclusively experimental in nature. The measurements are organised in two steps. Firstly, an $Ar/H_2/N_2/O_2$ mixture is used (i) to validate the sensitivity achieved for the OH detection, (ii) to find indications of plasma-surface interactions in our MW reactor, and (iii) to provide thereby a link to the results obtained in the ETP remote plasma. Secondly, the initial plasma chemical system is extended to a hydrocarbon containing discharge. IRLAS measurements on main stable and intermediate species in an $Ar/CH_4/N_2/O_2$ plasma are carried out. The results that are discussed in the context of this thesis represent only a fraction of a more extensive study [61] and focus essentially on the oxidation of the precursor. In particular the role of OH is discussed.

After a brief characterisation of the experimental arrangement and discharge parameters (section 3.2) the results are presented and discussed in section 3.3 following the two step approach. Conclusions are summarised in section 3.4.

3.2   Experimental

3.2.1   Discharge setup

The discharge was created in a planar microwave reactor which has been described in detail elsewhere [23,25,39]. For this reason only a brief account is given here. The experimental setup is shown schematically in figure 3.1.

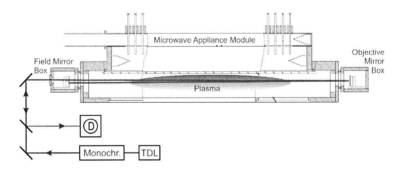

Figure 3.1:   Experimental arrangement of the planar microwave plasma reactor (side view) used in the IRLAS studies with a White type multiple pass optical arrangement, and tuneable diode laser (TDL) infrared source. The path of the diode laser beam is indicated (adapted from [25]).

The dimensions of the reactor vessel which was made from aluminium were $120 \times 21 \times 15 \text{ cm}^3$. The produced microwave radiation (Sairem GMP60KE/DC, $f = 2.45 \text{ GHz}$) was guided in a T-shaped appliance module and applied to the plasma volume through a quartz window. The input power was kept constant at 1.5 kW; the reflected power was typically less than 15 %. The active, relatively homogeneous plasma zone could be as long as ~ 100 cm and decayed over about 2 ...5 cm from the quartz window into the vessel volume. Typical electron densities (n_e) and electron temperatures (T_e) for this reactor are in the order of 10^{10} ... 10^{11} cm^{-3} and 1 ... 5 eV depending on the distance from the quartz window and coupling elements [62][1].

The reactor system was pumped down to 10^{-5} mbar in the standby mode by a turbomolecular pump backed by a rotary pump (Pfeiffer TMH64 and Leybold TriVac D40). During the measurements only the rotary pump was used. The pressure was kept constant at 1.5 mbar by means of a butterfly valve combined with a pressure gauge (MKS Baratron Type 127) and controller (MKS 600 series).

3.2.2   Injected precursors

Flowing gas conditions with a total gas flow rate F of 440 ... 460 standard cubic centimetres per minute (sccm)[2] were chosen for all experiments. The residence time of about 3 s follows from

$$\tau_{res} = \frac{pV}{p_0 F} \cdot \frac{T_0}{T_g},$$ (3 - 1)

where p, V denote reactor pressure and volume, p_0, T_0 are standard pressure and temperature (STP) (i.e., 1000 mbar, 273.15 K) and T_g is the gas temperature inside the reactor (here: 700 K, see 3.3). For both basic gas mixtures $H_2/N_2/O_2$ and $CH_4/N_2/O_2$ respectively an Ar background gas flow of 420 sccm was used resulting from an attempt to adjust the mean energy per feed gas particle E_{mean} between the present MW studies and the conditions of the ETP reactor (5 kW, 3000 sccm Ar, [49]). The mean energy may be introduced as

$$E_{mean} = \frac{P_{in}}{pV} k_B T_g \tau_{res} \sim \frac{P_{in}}{F},$$ (3 - 2)

with the injected power P_{in} and the Boltzmann constant k_B [3]. E_{mean} is proportional to the generalised (dimensionless) reactor parameter $\Gamma = E_{mean}/(k_B T_g)$ which can be defined for a plasma chemical system and compares E_{mean} of the feed gas particle with its thermal energy [63]. Similar values, among them the specific energy (P_{in}/F) or the Yasuda parameter ($P_{in}/(Fm_{molec})$) have been used for characterising reactive plasmas [64,65]. Theoretically, the present Ar flow should have been slightly higher. However the available pumping speed of the rotary pump limited the total gas flow to the present value since the total pressure of 1.5 mbar should be comparable to the studies of Hempel and Mechold [39,40]. It was estimated in [61] that under these experimental conditions the influence of charge transfer reactions between Ar$^+$ and injected feed gas molecules accompanied by their dissociation was considerably increased (compared to electron impact dissociation).

---

[1]   These values may be considered as an estimate since the reported values were obtained in a pulsed H_2 plasma at considerably higher input power. Nevertheless, a complimentary estimate in [61] yielded fair agreement.

[2]   Note: 60 sccm = 1 sccs (i.e., per second) and corresponds to 2.69×10^{19} molecules/s. 1 scc equals the number of molecules in 1 cm^3 at STP, i.e. (22 414)$^{-1}$ mol or 2.69×10^{19} molecules [65].

During the first series of measurements gas flows of 0 ... 20 sccm H_2 and 20 sccm $N_2 + O_2$ were admixed to the discharge. Thereby the N_2/O_2 ratio was varied. These conditions might be considered as similar to the studies at the ETP reactor where $H_2/N_2/O_2$ were admixed to the background, i.e. the low pressure vessel [49]. The second basic mixture was made of 10 sccm CH_4, 10 sccm N_2 and 0 ... 20 sccm O_2.

3.2.3   Spectroscopic issues

The microwave reactor being well-suited for end-on spectroscopic studies was equipped with a White type multiple pass cell optics [66] (figure 3.1) and adjusted to a total absorption path length (L_{eff}) of 60 m. The mirrors were separated by 1.5 m and provided an observation volume mainly corresponding to the active plasma zone. The multimode emission of the lead salt lasers (Laser Components) was collected by means of an off-axis parabolic mirror, spectrally filtered by a monochromator (Laser Analytics) and directed to the plasma chamber. Provided that sufficient optical power was available after the White cell a second monochromator was employed for filtering plasma induced noise superimposed on the optical signal which was finally detected by means of mercury cadmium telluride (MCT) detectors (Graseby Infrared, Judson J15 series). Different MCT detectors were chosen to match their high sensitivity regime to the (wide) spectral range applied here (500 ... 3000 cm^{-1} or 19 ... 3.3 μm).

Spectral scans were accomplished by means of the rapid scan approach [42]. The laser current was ramped, detector signals were acquired and analysed online by means of the *TDL Wintel* software package. Using this sweep integration method absolute molecular number densities from a fit of the recorded spectra to known spectral positions of the target species were obtained provided that discharge pressure p and gas temperature T_g are known. Additional details of the spectroscopic system can be found in [40]. Table 3.1 collects the detected molecules together with the spectral feature (position and line strength at the reference temperature) which was used to measure them in the present study. Except for C_2H_6, that required older line parameter data [67], the 2004 edition of the *HITRAN* database [68] was applied. Since the temperature dependence of the partition function $Q(T)$ is supplied as supplemental database material it is routinely considered by an approximation in *TDL Wintel*. The retrieved molecular concentrations were therefore corrected on the basis of the tabulated $Q(T)$ values (for more details see sub-sections 2.2.3 and 2.2.4). No error bars are given for the number densities in what follows. A discussion of accuracy and limitations is provided in chapter 7.

The detection of the hydroxyl radical in its fundamental band is challenging because interesting line positions coincide with the edge of the MIR spectral range, i.e. around 500 cm^{-1} (20 μm) or 3300 cm^{-1} (3 μm) where lead salt lasers and MCT detectors typically exhibit a limited performance. The OH concentrations retrieved here were obtained from two unresolved features at 532.14 cm^{-1} (table 3.1). Their lower level energy is relatively high which unambiguously supports measurements in the microwave plasma due to the increased gas temperature in this environment. The upper panel of figure 3.2 illustrates this fact: the theoretical transmission was calculated for a constant number density $n(OH) = 1.5 \times 10^{12}$ cm^{-3} at 300 K (plasma off) and 700 K (plasma on). In the latter case the temperature dependence of the line strength enables the OH feature to be observed and so is the lower panel. It should be mentioned that not only the performance of the TDL instrumentation degrades considerably to the edges of the MIR spectral range, but also the number of available references gases is limited to facilitate relative and absolute frequency calibration as well as a the check for monomode emission of the laser. The latter was achieved in this case by means of saturated absorption features of H_2O_2 being evaporated into the discharge chamber.

Molecule	Line Position [cm^{-1}]	Line Strength S [cm/molecule]
[a,b] NH_3	965.50	1.3×10^{-19}
[a,b] NO	[c] 1884.30	$\sim 2.8 \times 10^{-20}$
[a,b] N_2O	1308.39	8.9×10^{-20}
[a,b] H_2O	1884.57	2.0×10^{-21}
[a,b] OH	[c] 532.14	$\sim 2 \times 10^{-24}$
[b] CH_4	1327.07	9.6×10^{-20}
[b] C_2H_2	1307.16	1.3×10^{-19}
[b] C_2H_4	965.00	5.5×10^{-21}
[b] C_2H_6	[d] 2990.0x	$\sim 10^{-19}$
[b] HCN	1327.02	3.4×10^{-22}
[b] H_2CO	[c] 2802.77	$\sim 1.4 \times 10^{-19}$
[b] CO	2254.75	5.3×10^{-23}
[b] CO_2	664.20	4.3×10^{-20}

Table 3.1:
Overview of measured molecular species and the line strengths corresponding to the transitions that were used for their detection.

[a] detected in the H_2/N_2O_2 system
[b] detected in the CH_4/N_2O_2 system
[c] unresolved line; S is the sum of the individual transitions
[d] about 40 unresolved lines within 0.10 cm^{-1} at 2990.00 cm^{-1}; the straightforward sum over all transition yields $S \sim 4 \times 10^{-19}$ cm/molecule

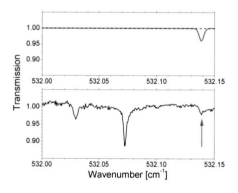

Figure 3.2:
Upper: Calculated transmission spectra corresponding to 1.5×10^{12} cm^{-3} OH for a gas temperature of 300 K (dotted black) and 700 K (solid grey). **Lower:** OH radical (indicated by an arrow) measured in an Ar/CH_4/O_2 plasma (420/10/20 sccm). An OCS reference gas cell was simultaneously placed in the beam path. Pressure and absorption path were $p = 1.5$ mbar, $L_{eff} = 60$ m in both cases.

3.3   Results and discussion

3.3.1   Discharge parameters

It has been established that for many (deposition) plasma chemical systems in general [63,65,69] and for hydrocarbon containing plasmas in particular [3,70] the Yasuda or reactor parameter and mean the energy E_{mean} (equation (3 - 2)) respectively provide an adequate means to generalise the behaviour the system, e.g. in respect to deposition rate, molecular concentrations etc. Considering all measurements that were carried out in the Ar/CH$_4$/(H$_2$)/N$_2$/O$_2$ model system [61] it transpires, however, that E_{mean} cannot be applied in the present study. In contrast to the basic idea that different measurements may be describe by a single generalised function of E_{mean}, the rescaled number densities would not yield such a result. Obviously a critical requirement for the application of the chemical quasi-equilibria approach is violated, i.e. $n_e \sim E_{mean}$ does not hold since the proportional factor shows a strong dependence on E/n [63]. This might be caused - amongst others - by the strong variation of the electron impact dissociation rate coefficients with E/n for this MW discharge [25].

A comparison of typical electron densities and temperatures between the MW and the ETP reactor shows that n_e is approximately in the same order of magnitude (10^{11} cm^{-3}) whereas the electron temperatures are clearly different [60,62]. For the ETP reactor T_e is about 0.3 eV. In other words, in contrast to the MW reactor ($T_e \geq 1$ eV), electron induced processes can be neglected.

In chapter 2 the importance of an adequate estimate of the gas temperature T_g for the determination of correct absolute molecular number densities was discussed (section 2.2.3, eq. (2 - 20)). In the present case the strong temperature dependence (i.e., the different lower state energies) of two H$_2$O lines at 1884.565 cm^{-1} and 1884.582 cm^{-1} respectively were used to extract T_g from spectral fits to the recorded spectra for different temperatures. A gas temperature of 700 K yielded best agreement and is used throughout further analysis. It should be considered as an average (line of sight) estimate.

3.3.2   (Ar -) H$_2$ - N$_2$ - O$_2$ discharges

As pointed out earlier the H$_2$ - N$_2$ - O$_2$ model system serves as to validate the sensitivity for the detection of the hydroxyl radical and to compare the results with those obtained in the ETP reactor where strong indications of surface production of the stable products were found. Figure 3.3 illustrates typical number densities measured in the MW reactor. The feed gases are efficiently converted into NH$_3$ (≤ 2 %) at very low oxygen contents or H$_2$O (25 %), NO (3.5 %) and N$_2$O (0.8 %) as soon as O$_2$ is added. The given abundances are expressed in effective mixing ratios, i.e. without taking into account the admixed Ar flow [61]. Similar to the ETP reactor a relatively high effective mixing ratio of a few per cent of NH$_3$ or NO is found in the MW reactor. It is even more interesting that the number density of ammonia decreases rapidly for small admixtures of O$_2$ to the discharge and falls below the detection limit for O$_2$ flows higher than 5 sccm (i.e., ≥ 20 % of the molecular precursors H$_2$/N$_2$/O$_2$). For all conditions where O$_2$ was present in the reactor the OH radical could be detected at relatively small number densities close to the detection limit which was $\sim 10^{11}$ cm^{-3} here. The effective mixing ratio for figure 3.3 would be 450 ppm.

In order to discuss the general trends of molecule formation the behaviour of ammonia, which was also extensively studied in the ETP reactor (without oxygen), will be first considered in more detail here. For this reason NH$_3$ concentrations normalised to their maximum are plotted in figure 3.4 for three different low pressure plasmas, among them a DC [51], the current MW and the expanding thermal plasma [53]. In the latter case, where a lot of different reactor geometries (surface/volume ratio of the vessel, arc configuration) were studied, a parameter set similar to the MW conditions has been chosen ($P_{in} \sim 3$ kW, 3000 sccm Ar flow through the arc and 300 sccm N$_2$ + H$_2$ mixture added in the background,

$p = 1$ mbar). For the DC discharge not all discharge parameters were available ($p = 2.66$ mbar, $N_2 - H_2$) [51]. Nevertheless, it can be concluded from figure 3.4 that the NH_3 formation follows a very similar behaviour in different discharges, particularly at low and high relative H_2 admixtures ($F(H_2)/F(H_2 + N_2)$) where the formation is limited by the number of available H or N radicals [47]. The maximum of NH_3 synthesis in the DC discharges is observed close to the stoichiometric case (75% H_2/25 % NH_3) whereas it is shifted to higher N_2 flows for the two other reactors.

The NH_3 example has been chosen since the surface dominated production of ammonia by subsequent pick-up reactions of H radicals from the gas phase by surface adsorbed NH_i ($i = 0 \ldots 2$) followed by NH_3 desorption is well-known and has been discussed and validated several times [53 and ref. therein, 55]. In this respect the very pronounced similarity between the MW and ETP configuration is surprising, because the reactor wall materials are clearly different (ETP: stainless steel, MW: aluminium). This suggests two conclusions:

i)   the molecule formation in the MW reactor (at least of NH_3) is also strongly influenced by surface reactions, and

ii)   the properties of the individual wall material are less important as long as a high flux of radicals, e.g. from a plasma are available leading to a hot or mobile surface layer [46].

The latter fact is additionally confirmed by experiments in the ETP reactor where the surface state was changed from stainless steel by depositing a-SiN_x and no changes in the NH_3 formation were found [53]. Further experiments revealed that the position of the maximum of NH_3 formation and the absolute number densities are mainly influenced by external discharge parameters, such as pressure, surface/volume ratio and the kind of discharge generation, i.e. the flow of produced radicals [52,53].

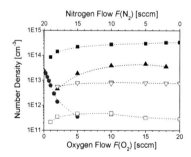
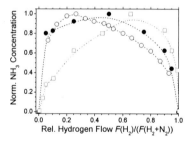

Figure 3.3:   Molecular number densities of H_2O (■), NH_3 (●), NO (▲), N_2O (▽), and OH (□) measured in an Ar/H_2/(N_2+O_2) (420/10/20 sccm) MW discharge (1.5 kW, 1.5 mbar). Dotted lines were added as a guide to the eye.

Figure 3.4: NH_3 concentrations (normalised to its maximum) observed in oxygen free $N_2 - H_2$ discharges: ● - MW reactor (1.5 mbar, 20 sccm $N_2 + H_2$ flow and 420 sccm Ar bath, 1.5 kW), O - ETP reactor (1 mbar, 300 sccm $N_2 + H_2$ background injected flow and 3000 sccm Ar through the arc, 55 A [53, fig. 5]), □ - DC glow discharge (2 Torr (2.66 mbar), 55 mA [51, fig. 3]).

Conclusion (i) for the MW reactor is further supported by studies on N_2/O_2 plasmas and additionally confirmed in the ETP reactor [48]. It has been established that N_2O is the dominant product at low O_2 admixtures while NO is the main stable species at higher oxygen contents. This has been explained by a surface covered with N atoms that are increasingly replaced by O atoms leading to the preferential surface production of NO and finally NO_2. The same trend can be detected in figure 3.3 for the MW discharge (the NO_2 density was still below the detection limit). The surface related production of NO showing similar trends has also been discussed and modelled by Gordiets et al. for a low pressure DC discharge [71].

By taking into account the strong influence of surface processes it becomes possible to explain the strong decrease of NH_3 with increasing O_2 flow (figure 3.3) qualitatively. This behaviour was additionally studied for different H_2 admixtures (5 ... 20 sccm) and is depicted in figure 3.5 in combination with the development of the OH density. It is evident from figure 3.5 that even higher H_2 flows cannot compensate the strong NH_3 decrease with O_2.

In equations (3 - 3) to (3 - 6) several gas phase reactions with a considerably high rate constant are summarised [56]. Specifically equations (3 - 3) and (3 - 4)) might explain the observed oxygen dependence of NH_3, i.e. ammonia and its precursors (NH_i) are affected by O and OH radicals in the gas phase. The system is characterised by a deficiency of precursor fragments for the ammonia production. On the other hand, e.g. the direct reaction $NH_2 + O$ was not considered as a preferential one amongst others in a kinetic study where OH appeared delayed as a product of a secondary reaction [72]. Moreover, all suggested gas phase reactions may not explain why an increase of the H_2 flow has almost no influence on the NH_3 mixing ratio.

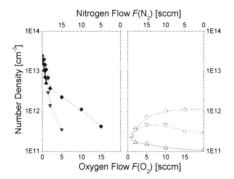

Figure 3.5:
Number densities of NH_3 (left) and OH (right) for different H_2 admixtures to an $Ar/(N_2+O_2)$ (420/20 sccm) MW discharge (1.5 kW, 1.5 mbar). The H_2 flows were 5 sccm (▲, △), 10 sccm (▼, ▽), and 20 sccm (◆, ◇) respectively. Dotted lines were added as a guide to the eye.

It is therefore reasonable to assume that surface reactions not only govern the formation of NH_3 in $H_2 - N_2$ and of NO in $N_2 - O_2$ mixtures, but also efficiently suppress the NH_3 formation in $H_2 - N_2 - O_2$ systems as soon as oxygen is added. As discussed above, the flux of O radicals lead to a change of the surface coverage. This enhances the formation of N_2O and NO at the expense of the ammonia production which in turn also reduces the subsequent dissociation of NH_3. The latter was found to be essential to provide precursors for the NH_i pick-up reactions at the surface [73]. Both effects are thus synergetic and lead to the observed strong decrease.

In contrast to ammonia the (effective) OH abundance increases with a higher H_2 content, namely from 180 ppm (5 sccm H_2) to 900 ppm (20 sccm H_2). For higher H_2 flows the OH maximum appears to be stoichiometric (figure 3.5) suggesting that equation (3 - 5) is dominant. (3 - 6) should be valid for both gas phase and the surface. The surface production of H_2O was discussed by Schram [46]. For higher H_2 and O_2 flows H_2O also serves as a precursor for OH via secondary reactions ($H_2O \rightarrow H_2O_2 \rightarrow OH$ [56]).

Preliminary modelling of the $Ar/H_2/N_2/O_2$ MW discharge, thereby only accounting for gas phase reactions and dividing the reactor into an active (topmost 3 cm) and a background part, underestimated the H_2O concentration by almost two orders of magnitude [74]. On the other hand, OH was overestimated by one order of magnitude. This underlines further the importance of surface reactions for molecule conversion.

$$NH_i + O \quad \rightarrow \quad NH_{i-1} + OH \ (i = 1,2,3) \qquad (3 - 3)$$
$$NH_3 + OH \quad \rightarrow \quad NH_2 + H_2O \qquad\qquad\quad (3 - 4)$$
$$H_2 + O_2 \quad \rightarrow \quad 2\,OH \qquad\qquad\qquad\quad (3 - 5)$$
$$H_2O + M \quad \leftrightarrow \quad OH + H + M \qquad\qquad (3 - 6)$$

3.3.3   (Ar -) CH_4 - N_2 - O_2 discharges

The previous model system was extended to an $Ar/CH_4/(H_2)/N_2/O_2$ plasma by adding methane. Typical number densities of main stable and intermediate species obtained while changing the O_2 flow (N_2 now constant) are shown in figure 3.6 and are summarised as effective mixing ratios (eMR) in table 3.2. Apart from the precursor CH_4 the detected species may be divided into 4 groups. The group of most abundant species (I) is formed by H_2O and CO as could be expected in an oxidising system. The next group of molecules (II) being present with more than 1 % effective mixing ratio consists of HCN, NH_3, NO and CO_2. Higher hydrocarbons (C_2H_y, group III) are only present to a lesser extent, especially the very low concentration of C_2H_6 is remarkable (see below). Finally, OH as a main radical and H_2CO as an early main intermediate molecule of the oxidising process [22] were detected with effective concentrations of almost 1000 ppm (group IV).

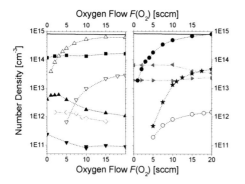

Figure 3.6:
Molecular number densities in an $Ar/CH_4/N_2/O_2$ (420/10/10/(0...20) sccm) MW discharge (1.5 kW, 1.5 mbar). The symbols represent ■ CH_4, △ CO, ▽ CO_2, ▲ C_2H_2, ▼ C_2H_6, ◇ H_2CO, ● H_2O, ◀ HCN, ▶ NH_3, ★ NO, ○ OH. The injected CH_4 density is indicated by a solid line. Dotted lines were added as a guide to the eye.

Apart from the fact (i) that the sum of the effective mixing ratios is > 1, a few other aspects will be discussed in what follows, namely (ii) the CO/CO_2 ratio, (iii) the low abundance of higher hydrocarbons, the behaviour of (iv) nitrogen containing products and (v) the hydroxyl radical.

Group I		Group II		Group III		Group IV	
Species	**eMR**	**Species**	**eMR**	**Species**	**eMR**	**Species**	**eMR**
CO	52 %	HCN	6.6 %	C_2H_2	0.2 %	H_2CO	0.09 %
H_2O	47 %	NH_3	1.7 %	C_2H_4 [a]	< 0.03 %	OH	0.08 %
CH_4	13.5 %	NO	1.3 %	C_2H_6	0.009 %		
		CO_2	1.0 %				

Table 3.2:   Summary of detected molecules and their effective mixing ratios (eMR) in an $Ar/CH_4/N_2/O_2$ MW discharge. Values were taken from figure 3.6 at $F(O_2) = 10$ sccm.

[a] mixing ratio was below detection limit; value estimated from line strength and noise level

The CH_4 containing plasma used here has a depletion of the precursor

$$D = 1 - \frac{n(CH_4)_{plasma}}{n(CH_4)_{in}},\tag{3-7}$$

of about 80 % which is almost constant for the studied oxygen flows. It is clearly higher than reported in [23,39] where the CH_4 depletion was about 30 % for a comparable O_2 content under flowing conditions. Although the total gas flow and the CH_4 flow was approximately the same in their measurements, they used only 1/7 of the present Ar flow, but an additional relatively high H_2 admixture. It can be assumed that the higher Ar content leads - as originally intended - to a better dissociation by charge transfer and dissociative recombination. Next, a fragmentation rate in molecules per J may be introduced as

$$R_f = \frac{F(CH_4)}{60} \cdot D \cdot \frac{n_0}{P_{in}},\tag{3-8}$$

where the precursor flow $F(CH_4)$ is given in sccm, n_0 $(= p_0/(k_B T_0))$ is the number density at STP and P_{in} is the injected power. R_f would be about 2.4×10^{15} J^{-1} and is thus comparable to an oxygen free $Ar/CH_4/H_2/N_2$ plasma studied by Hempel in the same MW reactor [40], even though the Ar flow was also only 1/7 of the currently applied bath gas flow.

Generally it can be concluded that the majority of molecular number densities that were observed in the present measurements are comparable to similar studies in the same MW reactor and an alternating current (AC) discharge [22,23] within one order of magnitude. It should be noted that the O_2 admixtures used here always correspond to low oxidising conditions in those studies where up to 500 sccm O_2 were employed [23]. Consequently the CO_2 selectivity, which may be defined as

$$S(CO_2) = \frac{n(CO_2)}{n(CO) + n(CO_2)},\tag{3-9}$$

is relatively small (i.e. less than 5 %) throughout the current experiments (figure 3.7, upper panel). This accords well with both earlier mentioned studies where $S(CO_2)$ showed values > 1 only in the case of the MW discharge with a high O_2 content (> 100 sccm) [22,23]. In contrast, $S(CO_2)$ > 1 was observed in almost all cases in a pure CH_4/O_2 RF discharges at very low total flow rates [18]. It is very likely that the longer residence time leads to a better (i.e. more efficient) oxidation to CO_2 as final product. This would also accord with the observations at low O_2 flows under static and flowing conditions by Mechold [39].

Next, the carbon balance B_C of the system was calculated on the basis of the number density n of measured C containing species

$$B_C = \frac{\sum_m m \times n(C_m H_x N_y O_z)}{n(CH_4)_{in}}$$
(3 - 10)

and is shown (in combination with individual contributions higher than 0.001) in the lower panel of figure 3.7. While at low O_2 flows up to 5 sccm the precursor and oxygen free products (C_2H_2, HCN) dominate B_C which is unambiguously lower than 100 %, the main contribution to B_C comes from CO at higher O_2 levels. In this case B_C > 1. The observed increase in the carbon balance strongly indicates a transition from a deposition into an etching system where a-C:H layers from the reactor walls are removed [61]. This agrees also with validation measurements carried out in the reactor after a longer break and a cleaning procedure. These measurements yielded much lower CO and HCN concentrations than before. An increase of B_C with oxygen was also observed in the earlier $Ar/CH_4/H_2/O_2$ MW studies by Mechold where a transition to B_C > 1 was detected at $O_2/CH_2 \sim 2:1$ [39]. Since H_2O is not included in the calculation of B_C, but also present with a high abundance at higher O_2 flows (figure 3.6), a sum of the effective mixing ratios eMR > 1 in table 3.2 is reasonable as soon as etching from a-C:H layers is additionally considered.

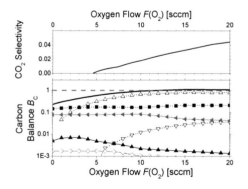

Figure 3.7:
Upper: CO_2 selectivity of the present $Ar/CH_4/N_2/O_2$ MW discharge. Lower: Carbon balance B_C (solid black) and main contributions (> 0.001) from the measured C containing species (see figure 3.6 for symbol definition). The theoretical maximum of $B_C = 1$ is indicated (dotted grey) Lines were added as a guide to the eye.

Higher hydrocarbons (C_2H_x) play only a minor role in the gas phase of the current measurements (table 3.2), i.e. the number densities of acetylene and, particularly, of ethane are lower (up to a few orders of magnitude) than observed in earlier experiments in the MW reactor, e.g. in the O_2 free system [25] as well as at low O_2 contents in [23]. This suggest an efficient quenching of C_2H_x via recombination of $CH_3 \rightarrow C_2H_6$ and subsequent H abstraction. It is interesting to note that NH_3 and HCN remain almost constant for the applied oxygen flows, i.e. the formation of NH_3 is - in contrast to the previous sub-section - not suppressed by O_2 addition. Since it was established in different H_2 - N_2 - O_2 systems that the NH_3 (N_2O and NO) formation strongly depends on the surface coverage with N and O atoms, the results of the methane containing plasma suggest a considerably different surface state. This hypothesis might be additionally supported by the observation that N_2O was readily detectable in an $Ar/N_2/O_2$ discharge, but as soon as traces of CH_4 were added the N_2O concentrations falls immediately below its detection limit. An excess of nitrogen in the system now does not automatically lead to the formation of nitrogen rich N_xO_y stable products as in 3.3.2. On the other hand NO exhibits an effective mixing ratio of > 1 % (table 3.2).

The number densities of H_2O and OH observed here accord very well with the earlier measurements or model calculations (OH) for this MW discharge [23,39]. Moreover, OH is in the same range as predicted by Fan et al. for the AC discharge [22]. Since both plasma chemical models were based on gas phase reactions, it might be concluded that the observed OH was produced via gas phase reactions whereas all other hydroxyl radicals being involved in surface reactions have such a short life time that they cannot contribute to the TDLAS signal. This is especially reasonable for the MW reactor geometry where TDLAS mainly observes the active zone of the plasma.

The main (gas phase) reaction path that may be extracted from the existing plasma chemical models [22,23] and support the current experimental data is given in equation (3 - 11).

$$CH_4 \xrightarrow{e,O} CH_3 \xrightarrow{O,OH} H_2CO \xrightarrow{OH} HCO \xrightarrow{O} CO \xrightarrow{O,OH} CO_2 \qquad (3 - 11)$$

The oxidation of CH_3 is a fast reaction [39] and hence efficiently suppresses the formation of higher hydrocarbons (C_2H_x), e.g. compared to oxygen free systems [25]. Formaldehyde is an important intermediate molecule of the early oxidation process. If the oxygen content of the system is low, the reaction path (3 - 11) already terminates at CO. This is confirmed by the low CO_2 selectivity (figure 3.7) and additionally supported by the decrease in H_2CO at O_2 flows higher than 10 sccm (figure 3.6). The main production and depletion reactions of CO (i.e. produced via HCO and lost into CO_2) were also confirmed in a model calculation for the already mentioned RF discharge [18,19].

3.4   Summary and conclusions

TDLAS measurements in the MIR spectral range were performed in two different plasma chemical model systems, i.e. $(Ar)/H_2/N_2/O_2$ and $(Ar)/CH_4/N_2/O_2$. The focus of the current investigations in a planar microwave discharge reactor encompassing the detection of up to 13 different molecular species was on the detection of the hydroxyl radical using a conventional *in-situ* multiple pass cell arrangement. Additionally, indications of surface reactions and their influence on the molecule conversion in the MW reactor, which has been treated theoretically only by pure gas phase chemistry so far, were analysed. The experimental conditions were adapted to measurements in a remote ETP plasma by choosing a high Ar flow in order to enhance charge exchange reactions and dissociative recombination of the precursors. Similar to the results in the recombining ETP plasma (and other reactors) indications of the importance of plasma surface interactions for the formation of the most abundant molecules NH_3 and NO were found in the planar MW reactor. Particularly, the strong decrease of NH_3 due to the addition of oxygen to the system may be described by the change of the composition of a mobile surface layer where N atoms are increasingly substituted by O. This plasma produced surface layer also "hides" the original properties of the reactor wall material (e.g. stainless steel or aluminium). The mismatch of H_2O and OH concentrations between experimental data (OH $\sim 10^{11}$ cm^{-3}) and a preliminary (gas phase) model of the H_2 - N_2 - O_2 system underlines the non-negligible role of the surface for molecule conversion and abundance of stable products in low pressure plasmas. Nevertheless, the present study confirms that measuring and analysing stable species instead of the usually more complicated detection of radicals can already reveal valuable information about plasma chemistry and plasma surface interactions.

In contrast to the H_2 - N_2 - O_2 system, the predicted OH number densities for methane containing plasmas (10^{12} cm^{-3}) could be confirmed, though the earlier developed models only accounted for gas phase reactions. This suggests that the hydroxyl radicals, being observed for the first time in reactive plasmas in the MIR, represent the active zone of the plasma. The present experimental conditions are characterised by an O_2 deficiency which is generally the case in air plasmas, e.g. for removal of volatile organic compounds. Consequently, an incomplete oxidation leading to a low CO_2 selectivity and relatively high formaldehyde concentrations was observed. Furthermore a transition from a deposition to an etching system was found resulting in a carbon balance higher than 100 % for O_2/CH_4 ratios > 1:1. The abundance of the main stable products could be calculated by means of a preliminary model of the reactor consisting of an active and a background part where additional C and H sources were introduced to account for the etching mode [74,75]. It transpires, however, that number densities of additional major intermediate species or radicals, specifically HCO, are substantial for the assessment of the oxidation and surface processes.

Further discrimination of surface and gas phase reactions can be accomplished by time resolved measurements. This encompasses both investigations of catalytic surface reactions on a s und sub-s time scale in the afterglow [49] or measurements in the sub-ms range yielding information on the plasma kinetics (chapter 5). The latter has become feasible with the advent of quantum cascade lasers (QCLs). However, different operation modes require first a thorough assessment of their application to low pressure (plasma) conditions (chapter 4). Employing QCLs in combination with optical resonators (chapter 6) may also tackle another shortcoming in the present TDLAS study, namely the sensitivity being limited to about 10^{11} cm^{-3} (see chapter 7 for a detailed discussion). This limit of detection is usually not as low as required for detecting highly reactive radicals with short life times (OH, C_2H, etc.). Even the observed OH densities were close to the detection limit in a spectral range where TDLs show a limited performance. Particularly in the emerging field of atmospheric pressure discharges or microplasmas where the interaction length between laser and plasma volume is inherently limited alternative approaches for measuring reactive radicals (such as OH) are highly desirable.

Bibliography

[1]   M. Weiler, R. Kleber, K. Jung, H. Erhardt, *Diam. Rel. Mater.* **1**, 121 (1992).

[2]   M. Bauer, T. Schwarz-Selinger, H. Kang, A. von Keudell, *Plasma Sources Sci. Technol.* **14**, 543 (2005).

[3]   A. von Keudell, I. Kim, A. Consoli, M. Schulze, A. Yanguas-Gil, J. Benedikt, *Plasma Sources Sci. Technol.* **16**, S94 (2007).

[4]   K. Matyash, R. Schneider, A. Bergmann, W. Jacob, U. Fantz, P. Pecher, *J. Nucl. Mater.* **313–316**, 434 (2003).

[5]   H.T. Do, V. Sushkov, R. Hippler, *New Journ. Phys.* **11**, 033020 (2009) [and references therein].

[6]   Y. Yang, Ph.D. Thesis, University Greifswald, 2001.

[7]   T. Lang, Ph.D. Thesis, University Vienna (TU Wien), 1993.

[8]   F.J. Gordillo-Vazquez, C. Gomez-Aleixandre, J.M Albella, *Plasma Sources Sci. Technol.* **10**, 99 (2001).

[9]   J. Ma, M.N.R. Ashfold, Y.A. Mankelevich, *J. Appl. Phys.* **105**, 043302 (2009).

[10]   K. Hassouni, X. Duten, A. Rousseau, A. Gicquel, *Plasma Sources Sci. Technol.* **10**, 61 (2001).

[11]   J. Hirmke, F. Hempel, G.D. Stancu, J. Röpcke, S.M. Rosiwal, R.F. Singer, *Vacuum* **80**, 967 (2006).

[12]   J. Hirmke, A. Glaser, F. Hempel, G.D. Stancu, J. Röpcke, S.M. Rosiwal, R.F. Singer, *Vacuum* **81**, 619 (2007).

[13]   L. Mechold, J. Röpcke, X. Duten, A. Rousseau, *Plasma Sources Sci. Technol.* **10**, 52 (2001).

[14]   J. Röpcke, L. Mechold, X. Duten, A. Rousseau, *J. Phys. D: Appl. Phys.* **34**, 2336 (2001).

[15]   G. Lombardi, K. Hassouni, G.D. Stancu, L. Mechold, J. Röpcke, A. Gicquel, *Plasma Sources Sci. Technol.* **14**, 440 (2005).

[16]   G. Lombardi, K. Hassouni, G.D. Stancu, L. Mechold, J. Röpcke, A. Gicquel, *J. Appl. Phys.* **98**, 053303 (2005).

[17]   J.R. Fincke, R.P. Anderson, T. Hyde, B.A. Detering, R. Wright, R L. Bewley, D.C. Haggard, W.D. Swank, *Plasma Chem. Plasma Process.* **22**, 105 (2002) [and references therein].

[18]   C. Busch, I. Möller, H. Soltwisch, *Plasma Sources Sci. Technol.* **10**, 250 (2001).

[19]   I. Möller, A. Serdyuchenko, H. Soltwisch, *J. Appl. Phys.* **100**, 033302 (2006).

[20]   R. Beckmann, W. Kulisch, H.J. Frenck, R. Kassing, *Diam. Rel. Mater.* **1**, 164 (1992).

[21]   Y. Liou, Y.R. Ma, *Diam. Rel. Mater.* **3**, 573 (1994).

[22]   W.Y. Fan, P.F. Knewstubb, M. Käning, L. Mechold, J. Röpcke, P.B. Davies, *J. Phys. Chem. A* **103**, 4118 (1999).

[23]   J. Röpcke, L. Mechold, M. Käning, W.Y. Fan, P.B. Davies, *Plasma Chem. Plasma Process.* **19**, 395 (1999).

[24]   P.K. Bachmann, D. Leers, H. Lydtin, *Diam. Rel. Mater.* **1**, 1 (1991).

[25]   F. Hempel, P.B. Davies, D. Loffhagen, L. Mechold, J. Röpcke, *Plasma Sources Sci. Technol.* **12**, S98 (2003).

[26]   C.D Pintassilgo, G. Cernogora, J. Loureiro, *Plasma Sources Sci. Technol.* **10**, 147 (2001).

[27]   M. Kareev, M. Sablier, T. Fujii, *J. Phys. Chem. A* **104**, 7218 (2000).

[28]   J.C. Legrand, A.M. Diamy, R. Hrach, V. Hrachova, *Vacuum* **50**, 491 (1998).

[29]   F.L. Tabares, V. Rohde, ASDEX Upgrade Team, *Plasma Phys. Control. Fusion* **46**, B381 (2004).

[30]   B.M. Penetrante, R.M. Brusasco, B.T. Merritt, G. E. Vogtlin, *Pure Appl. Chem.* **71**, 1829 (1999).

[31]   B.M. Penetrante, M.C. Hsiao, J.N. Bardsley, B.T. Merritt, G.E. Vogtlin, A. Kuthi, C.P. Burkhart, J.R. Bayless, *Plasma Sources Sci. Technol.* **6**, 251 (1997).

[32]   A. Ogata, K. Mizuno, S. Kushiyama, T. Yamamoto, *Plasma Chem. Plasma Process.* **18**, 363 (1998).

[33]   A. Ogata, K. Yamanouchi, K. Mizuno, S. Kushiyama, T. Yamamoto, *Plasma Chem. Plasma Process.* **19**, 383 (1999).

[34]   F. Thevenet, O. Guaitella, E. Puzenat, C. Guillard, A. Rousseau, *Appl Cat. B: Environm.* **84**, 813 (2008).

[35]   Z. Machala, M. Janda, K. Hensel, I. Jedlovsky, L. Lestinska, V. Foltin, V. Martisovits, M. Morvova, *J. Mol. Spectrosc.* **243**, 194 (2007).

[36]   A.V. Pipa, T. Bindemann, R. Foest, E. Kindel, J. Röpcke, K.D. Weltmann, *J. Phys. D: Appl. Phys.* **41**, 194011 (2008).

[37]   A.V. Pipa, J. Röpcke, *IEEE Trans. Plasma Sci.*, ISSN 0093-3813, accepted (2009).

[38]   T. Schwarz-Selinger, R. Preuss, V. Dose, W. von der Linden, *J. Mass Spectrom.* **36**, 866 (2001).

[39]   L. Mechold, Ph.D. Thesis, University Greifswald, ISBN 978-3-89722-687-6, 2001.

[40]   F. Hempel, Ph.D. Thesis, University Greifswald, ISBN 978-3-8325-0262-1, 2003.

[41]   R. Hemberger, S. Muris, K.U. Pleban, J. Wolfrum, *Combust. Flame* **99**, 660 (1994).

[42]   D.D. Nelson, M.S. Zahniser, J.B. McManus, C.E. Kolb, J.L. Jimenez, *Appl. Phys. B* **67**, 433 (1998).

[43]   X. Mercier, P. Jamette, J.F. Pauwels, P. Desgroux, *Chem. Phys. Lett.* **305**, 334 (1999).

[44]   Z.W. Liu, X.F. Yang, A.M. Zhu, G.L. Zhao, Y. Xu, *Eur. Phys. J. D* **48**, 365 (2008).

[45]   R. Peeters, G. Berden, G. Meijer, *Appl. Phys. B* **73**, 65 (2001).

[46]   D.C. Schram, *Plasma Sources Sci. Technol.* **18**, 014003 (2009).

[47]   D.C. Schram, R.A.B. Zijlmans, J.H. van Helden, O. Gabriel, G. Yagci, S. Welzel, J. Röpcke, R. Engeln, *Journ. Optoelectronics Adv. Mater.* **10**, 1904 (2008).

[48]   J.H. van Helden, Ph.D. Thesis, Eindhoven University of Technology, ISBN 978-90-386-2511-9, 2006.

[49]   R.A.B. Zijlmans, Ph.D. Thesis, Eindhoven University of Technology, ISBN 978-90-386-1288-1, 2008.

[50]   E.N. Eremin, A.N. Maltsev, V.L. Syaduk, *Russ. Journ. Phys. Chem.* **45**, 635 (1971).

[51]   J. Amorim, G. Baravian, G. Sultan, *Appl. Phys. Lett.* **68**, 1915 (1996).

[52]   P. Vankan, T. Rutten, S. Mazouffre, D.C. Schram, R. Engeln, *Appl. Phys. Lett.* **81**, 418 (2002).

[53]   J.H. van Helden, W. Wagemans, G. Yagci, R.A.B. Zijlmans, D.C. Schram, R. Engeln, G. Lombardi, G. D. Stancu, J. Röpcke, *J. Appl. Phys.* **101**, 043305 (2007).

[54]   B. Gordiets, C.M. Ferreira, M.J. Pinheiro, A. Ricard, *Plasma Sources Sci. Technol.* **7**, 363 (1998).

[55]   B. Gordiets, C.M. Ferreira, M.J. Pinheiro, A. Ricard, *Plasma Sources Sci. Technol.* **7**, 379 (1998).

[56]   M. Capitelli, C.M. Ferreira, B.F. Gordiets, A.I. Osipov, *Plasma Kinetics in Atmospheric Gases*, ISBN 3-540-67416-0, (Springer, Berlin, 2000).

[57]   J.H. van Helden, R.A.B. Zijlmans, D.C. Schram. R. Engeln, *Plasma Sources Sci. Technol.* **18**, 025020 (2009).

[58]   M.C.M. van de Sanden, R.J. Severens, W.M.M. Kessels, R.F.G. Meulenbroeks, D.C. Schram, *J. Appl. Phys.* **84**, 2426 (1998).
        M.C.M. van de Sanden, R.J. Severens, W.M.M. Kessels, R.F.G. Meulenbroeks, D.C. Schram, *J. Appl. Phys.* **85**, 1243 (1999).

[59]   R. Engeln, S. Mazouffre, P. Vankan, I. Bakker, D.C. Schram, *Plasma Sources Sci. Technol.* **11**, A100 (2002).

[60]   P.J. van den Oever, J.L. van Hemmen, J.H. van Helden, D.C. Schram, R. Engeln, M.C.M. van de Sanden, W.M.M. Kessels, *Plasma Sources Sci. Technol.* **15**, 546 (2006).

[61]   R.A.B Zijlmans, O. Gabriel, S. Welzel, F. Hempel, J. Röpcke, R. Engeln, D.C. Schram, *Plasma Sources Sci. Technol.* **15**, 564 (2006).

[62]   A. Rousseau, E. Teboul, N. Lang, M. Hannemann, J. Röpcke, *J. Appl. Phys.* **92**, 3463 (2002).

[63]   F. Miethke, Ph.D. Thesis, University Greifswald, 1996.

[64]   A. Rutscher, H.E. Wagner, *Plasma Sources Sci. Technol.* **2**, 278 (1993).

[65]   H. Yasuda, *Plasma Polymerization*, ISBN 0-12-768760-2, (Academic Press, London, 1985).

[66]   J.U. White, *J. Opt. Soc. Am.* **32**, 285 (1942).

[67]   L.S. Rothman *et al.*, *J. Quant. Spectrosc. Radiat. Transfer* **82**, 5 (2003).

[68]   L.S. Rothman *et al.*, *J. Quant. Spectrosc. Radiat. Transfer* **96**, 139 (2005).

[69]   K. Li, O. Gabriel, J. Meichsner, *J. Phys. D: Appl. Phys.* **37**, 588 (2004).

[70]   M. Bauer, Ph.D. Thesis, University Bayreuth, 2004.

[71]   B. Gordiets, C.M. Ferreira, J. Nahorny, D. Pagnon, M. Touzeau, M. Vialle, *J. Phys. D: Appl. Phys.* **29**, 1021 (1996).

[72]   J.D. Adamson, S.K. Farhat, C.L. Morter, G.P. Glass, R.F. Curl, L.F. Phillips, *J. Phys. Chem.* **98**, 5665 (1994).

[73]   J.H. van Helden, P.J. van den Oever, W.M.M. Kessels, M.C.M. van de Sanden, D.C. Schram, R. Engeln, *J. Phys. Chem. A* **111**, 11460 (2007).

[74]   R. Zijlmans (*private communication*).

[75]   W.J. Engelen, *Modeling of the Molecule Synthesis in an Ar-CH$_4$-N$_2$-O$_2$ Microwave Plasma*, Report, Eindhoven University of Technology, 2008.

4   Quantum cascade laser absorption spectroscopy

4.1   Introduction and motivation

The methods of absorption spectroscopy are of great importance in physics and chemistry because they provide a means of species identification and quantification, particularly in their ground states. Laser sources have the advantage of high spectral intensity, tuneability and narrow spectral bandwidth. The latter fact is essential for gas phase spectroscopy where narrow absorption features are studied. Measuring molecular species usually requires light sources in the infrared spectral range, specifically in the mid infrared (MIR) region, where the fundamental, i.e. strongest ro-vibrational, absorption features are situated. However infrared laser absorption spectroscopy (IRLAS) remained a diagnostic tool for niche applications, mainly for laboratory research, since suitable tuneable, light sources in the MIR range were often bulky and restricted to specific wavelength regions (e.g., CO_2, CO lasers), of low output power (DFG, OPO systems) or required cryogenic cooling, such as lead salt diode lasers (see chapter 2). In particular, in the field of semiconductor based lasers this situation has changed with the advent of quantum cascade lasers (QCLs).

This new class of lasers offers several advantages over tuneable diode lasers (TDLs), among them room temperature operation, increased radiation power and mode-hop free tuneability. The emission wavelength of QCLs is no longer restricted by the band gap energy and hence the composition of lead salts. Modern band-gap engineering techniques offer custom tailored emission beyond 3 µm. The basic concept is based on resonant tunneling between confined quantum states (chapter 2). Progress in band-gap engineering and thin film deposition techniques led to the development of the first QCL in 1994 and, finally, to their commercial availability. In the past decade the design and efficiency of QCLs has been improved by incorporating a DFB grating into the laser structure, yielding continuously tuneable single mode emission, and by optimising the active zone design and thermal management. QCLs are now considered as a valuable alternative light source for the MIR spectral region. While the first laser devices could be used at room temperature only in pulsed mode at low duty cycles, room temperature continuous wave (cw) QCLs have only recently been available.

Thus, in the first spectroscopic application of a QCL a pulsed laser has been employed being swept in its frequency by impressing a current ramp on a train of pulses [1]. Alternatively, single longer pulses were applied [2,3]. While the first spectrometers based on QCLs clearly focussed on gas phase spectroscopy, i.e. trace gas measurements [1,4-6], the further development have led to a diversified situation. Trace gas sensing now encompasses isotope measurements [7-10], atmospheric sensing [11-15], detection of explosives [16-18], liquid chromatography [19] or breath analysis [20,21] and enables highly sensitive measurements to be performed [22,23]. Recently, the application has been extended to plasma diagnostics [24-31]. The spectral characteristics of QCLs further facilitated studies of non-linear absorption phenomena, such as Lamb-Dip spectroscopy [32,33], fast passage effects or coherent population transfer [34,35], and the application of optical cavities [36,37][1] which was recently extended to the field of frequency metrology [38,39]. Since the narrow total

---

[1]   A detailed introduction to the combination of QCLs and optical resonators can be found in chapter 6.

tuning range of standard devices is a potential drawback for broadband absorption features, external cavity lasers (ECLs) covering 50 ... 100 cm^{-1} have been developed and employed specifically for measuring explosives, liquid samples. [17,35,40-43]. The ECL approach thereby also opens up the possibility of spectral imaging in the MIR [43]. Additionally, QCLs were tested for non-spectroscopic techniques such as free-space optical transmission of data [44-47] and near field microscopy [48], respectively. The intersubband transitions of QCLs can be also designed to yield THz emission (> 20 μm) which will be not considered in this thesis.

Early experiments with pulsed QCLs combined short laser pulses with the conventional method for sweeping lead salt lasers in tuneable diode laser absorption spectroscopy (TDLAS) by ramping a DC current [1,49,50]. Extensions originally developed for TDLAS, e.g., the sweep integration method [51,52], were adapted to pulsed QCL spectrometers [22]. Later, the inherent frequency chirp of the lasers was exploited by using long pulses to acquire absorption spectra during the pulse [2,3]. Although the successful implementation of QCLAS was very often reported in the literature, obstacles in the spectra, i.e. increased laser line widths along with pulsed QCLs [4] or asymmetric line shapes in both the short pulse [22,53,54] and the long pulse mode [55], have been recognised in several studies. Additionally, the latter method suffers from non-linear absorption effects [56]. The majority of these issues are not known from conventional TDLAS and degrade the performance of QCL absorption spectroscopy (QCLAS) spectrometers.

At low pressures the laser line width of pulsed QCLs was found to exceed the typical Doppler broadening [4] due to the frequency chirp and non-linear effects (e.g., fast passage) appeared more pronounced [56]. A few studies have been concerned about these drawbacks, however, a consistent explanation is absent so far, although all obstacles seem to be connected with the chirped laser. While a theoretical description for the non-linear absorption phenomena based on optical Bloch equations exist [34,55,56], the origin of distorted line shapes in the short pulse mode is not yet completely understood [22]. The problem is often empirically minimised by finding a compromise between a reasonable SNR and a narrow spectral width [57]. The sensitivity of spectrometers or quantitative results are typically verified by calibrating an effective laser line width or correction factors are abstracted by comparing measurements and simulated spectra [e.g., 22,50,58]. Only in a few studies the results were validated with independent counter-measurements by means of tuneable diode lasers (TDLs) [22,53] or cw QCLs [54,57]. Surprisingly and in contrast to the majority of other long pulse spectra under low pressure conditions, indications for a fast passage were not observed in such a comparative study by Grouiez et al. employing both a pulsed and a cw laser [54]. Recently, an intercomparison between the short and long pulse method has been in the centre of interest [59]. Unfortunately, this study was limited to pressure ranges (hundreds of mbar) where the artefacts diminish or disappear. A comparison of QCL tuning methods at lower pressures led to the conclusion that intermediate pulse lengths should be used to achiever better results [54].

Forthcoming applications of pulsed QCLs for, particularly, low pressure plasma diagnostics comprise therefore both desirable properties of the lasers such as mW radiation power obtained near room temperature as well as pulsed operation enabling a rapid frequency tuning and thus high time resolution, and not well-understood artefacts in the recorded spectra. A thorough and consistent comparison of main tuning methods for pulsed QCLs is still absent for a wider pressure range and complicates the decision for proper operation conditions. Meanwhile cw QCLs can provide a much better system performance than pulsed devices [57,60,and chapter 6] and may extend the precision of MIR spectrometers with the increasing availability of near room temperature cw lasers [60]. Nevertheless, pulsed QCLs appear perfectly suited for several spectroscopic applications, e.g., CRDS, where the incident radiation has to be interrupted to observe the light decay, or highly time resolved measurements exploiting the frequency chirp of lasers as already suggested in the late 1970s for lead salts [61].

This thesis work especially concerns the latter aspects and thus this chapter focuses (i) on a detailed analysis of appearance, origin and effects of the reported obstacles and (ii) on deducing guidelines for low pressure plasma diagnostics. At first, the characteristics of common tuning methods are established and compared with a co-aligned TDLAS system (section 4.2). The measurements start at elevated pressures where typically less difficulties are expected and are extended to low pressure conditions. Since the time consuming numerical calculation of non-linear absorption phenomena hampers an implementation in the data analysis procedure, alternative approximations are evaluated. In section 4.3 the properties of chirped QCL pulses are examined. A time resolved analysis of single chirped pulses in the short pulse mode is presented in section 4.4 and reveals a complex combination of three key factors leading to apparently increased effective laser line widths for pulsed QCLs. The present measurements combine and extend earlier approaches studying this effect [22,54]. Almost all experiments were carried out with the same pulsed laser to facilitate a fair comparison and exclude the influence of the device performance on the results. Finally, the consequences of the individual experiments are summarised (section 4.5). The results can be generalised and lead to a sound picture comprising both short and long laser pulses. Additionally, guidelines for QCLAS at low pressures are given.

4.2   Comparison between QCLAS and TDLAS

In recent years sophisticated and powerful sweep integration methods have been developed and combined with TDL spectrometers [62,63] enabling a real-time analysis of the acquired spectra. Since QCLs are often considered as substitutes to TDLs in the MIR [64], adapting these conventional principles and setups appears obvious. Similar to lead salt lasers the frequency sweep of QCLs is based on the change of the refractive index of the gain medium [65,66]. While this is temperature and charge carrier density dependent for TDLs, QCLs can only be tuned by temperature changes. Since the required input power for QCLs is considerably higher than for lead salts an indirect temperature sweep is nevertheless possible by means of a current modulation. The pulsed operation of QCL facilitates additional methods for sweeping the laser frequency which are briefly discussed in section 4.2.1. The experimental arrangement comprising a QCLAS and a co-aligned TDLAS setup is presented in 4.2.2. A comparison of both methods at elevated (section 4.2.3) and low pressure conditions (section 4.2.4) follows.

4.2.1   Modes of operation

A straightforward and direct but relatively slow method of tuning the QCL is the modulation of its heat sink temperature [49,58,67], e.g. with an external voltage ramp applied to the thermoelectric cooler (TEC) of the QCL (figure 4.1 a). In what follows this method is referred to as TEC modulation. The laser is operated with intermediate pulse lengths of a few tens of ns and the emission is detected with a high bandwidth thermoelectrically (TE) cooled detector and a fast digitising oscilloscope. Monitoring the signal at a constant temporal position during the laser pulse excludes the influence of the frequency chirped QCL pulse[2]. Due to the millisecond response time of the TEC the modulation causes a smooth frequency sweep between the subsequent laser pulses and a full scan may require several seconds. Depending on the characteristics of the laser and the TEC controller the external modulation voltage may be gradually increased or suddenly switched between its maximum and minimum (cf. triangular shape in figure 4.1 a). In the latter case the TEC controller governs the tuning. Choosing the first option might cause additional noise due to oscillations of the controller and thus the laser temperature.

Since the acquisition of a single spectrum requires up to tens of seconds this technique is not appropriate for monitoring rapidly changing molecular concentrations. However it enables extended spectral scans up to the almost entire emission range of the DFB QCL being typically less than 7 cm^{-1} (chapter 2). This is of particular interest for broad complex molecular absorption features where a wider scan range is preferable. Other tuning options does not provide a similar total sweep range. Recently, QCL devices on diamond based active sub-mounts with sophisticated resistive heaters included in the heat sink have been developed [68]. These lasers can be tuned over ~ 30 cm^{-1} from cryogenic to near room temperature on a sub-second scale. [68,69)

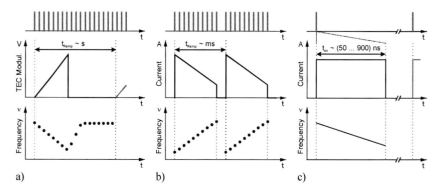

Figure 4.1:   Schematic diagrams of selected QCL tuning mechanisms by means of a (a) modulation of the TEC controller, (b) continuous current ramp superimposed on a train of short laser pulses or (c) single, long pulse with inherent tuning. Black circles represent the swept emission frequency of each laser pulse (represented by vertical lines).

A continuous DC ramp superimposed on a train of short laser pulses [1,22,49,70-72] can also accomplish a temperature sweep. This is frequently referred to as the *inter* pulse mode, i.e. several A are provided by triggered current pulses overlaid with a sub-threshold current ramp. This causes a variable thermal load and tunes the emission frequency (figure 4.1 b). The pulses vary from a few to up to tens of ns with a duty cycle of considerably

---

[2]   More details on this approach can be found in section 4.4.

less than 5 %. The current is ramped over several hundred pulses. Each laser pulse represents a spectral data point in the spectrum. A full spectrum is therefore acquired in ms (figure 4.1 b). The main challenges of this method are pulse-to-pulse amplitude and frequency fluctuations inherent to the short laser pulses [23].

This tuning method was adapted from conventional TDLAS and appears to be a straightforward replacement of the MIR light source while employing the same arrangements and analysis techniques with ms time resolution [22,23]. Later in this chapter (section 4.4) a detailed discussion shows that this approach is only partly useful. Furthermore the maximum average current through the laser sets the limit for the current ramp and thus for the total frequency sweep of the QCL being typically less than 1 cm^{-1}. The recent progress in designing QCL devices with improved heat removal from the active zone [73-77], as a condition for cw operation, has led to a reduced tuning efficiency with this approach.

Another tuning option yielding entire spectra of ~ 0.5 - 2 cm^{-1} per pulse is the *intra* pulse mode, i.e. the scanning in single, longer pulses [2,3]. The inherent heating of the laser due to current pulses of a few A results in a frequency-down chirp (figure 4.1 c). The chirp rate is mainly determined by the material and design of the device and cannot be controlled independently from other parameters as in the above mentioned methods, e.g., adjusting the output power with the seed current also influences the internal heating and thus the chirp rate. Since a single laser pulse provides a full spectrum highly time resolved measurements down to the pulse length are possible for quantitative measurements of rapidly changing chemical processes. Fast and high bandwidth detectors and digitising electronics are required because single absorption lines are scanned during 1 - 2 ns. Furthermore the rapidly chirped laser pulse can cause non-linear absorption behaviour (i.e., fast passage and power saturation effects, section 4.3).

Although not considered here, where the most common approaches are in the centre of interest, the pulsed operation of QCLs provides the option to modulate the duty cycle. Sweeping the pulse length or repetition frequency also varies the temperature balance of the QCL and rapid laser scans become possible [Ref:aphb_75p351_kosterev]. The sweep range constraints with current devices and the required time for a scan are comparable to the *inter* pulse mode, but at a reduced thermal load because no extra current is fed to the laser.

4.2.2   Experimental arrangement

The QCLs in the present experiments were driven with a commercial QCL measurement and control system (Q-MACS, neoplas control) and operated in a temperature stabilised housing. For all operation modes the lasers were triggered using a computer controlled data acquisition card (DAC) (National Instruments, PCI-6110E). The QCL pulses were detected with TE cooled detectors encompassing high bandwidth preamplifiers (neoplas control, VIGO elements) and a fast digitising oscilloscope (LeCroy WR104Xi). Except for the inherent tuning in the *intra* pulse mode the lasers had to be frequency scanned either by an externally ramped heat sink temperature for TEC modulation or by means of a current ramp for the *inter* pulse mode. This was generated from an externally supplied voltage ramp provided by the software package *TDL Wintel* in combination with a DAC. The program sweeps the laser, acquires the absorption spectra and analyses them simultaneously [22]. A short gate-off period at the end of each sweep suppresses the laser emission for measuring the detector offset voltage. The details of the presently used data acquisition and laser synchronisation principle for the *inter* pulse mode are essentially the same as described by Nelson et al. [23] except for the QCL driver (Q-MACS instead of the original Alpes Lasers Starter Kit).

The radiation of the lasers was firstly collected with an off-axis parabolic (OAP) mirror of 50 mm diameter (f/1.0) and guided to an aluminium vacuum vessel (~ 45 l) equipped with multiple pass cell optics. The mirrors were separated by 1.5 m and arranged in White cell configuration [79]. After 16 passes through the vessel the laser beam was further directed to an OAP of 25.4 mm diameter (f/0.64) in front of the detector. In order to compare the results from QCLAS and TDLAS measurements under identical conditions a TDL system (Infrared multi-component acquisition system - IRMA [62]) was co-aligned to the described optical arrangement (figure 4.2). The lead salt laser beam could be directed to the multi-pass cell with a retractable mirror. The TDL signal was then detected with a liquid nitrogen (LN) cooled detector (Graseby Infrared) which was inserted into the beam path by means of another retractable mirror. The TDLs were driven with a current ramp provided by a second *TDL Wintel* based system independently working from the QCLAS setup.

The vacuum cell was equipped with two capacitance gauges (MKS) of complementary measurement ranges enabling experiments between 0.1 mbar and 200 mbar. During the experiments the cell was pumped by a rotary pump with its port located at one end of the vessel whereas the gas input and pressure gauges were mounted at the opposite site. This configuration provided a homogeneous gas distribution over the entire base length of the multi-pass cell. Below 10 mbar the pressure was controlled by means of a butterfly valve in combination with one of the pressure gauges, i.e. the experiments could be carried out under flowing gas conditions at constant pressure. At elevated pressures the spectra had to be acquired under static gas conditions.

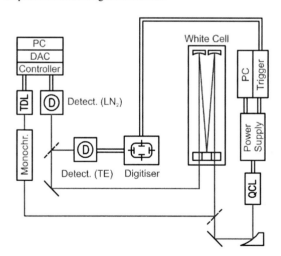

Figure 4.2:
Dual laser arrangement for comparative measurements between QCLAS and TDLAS. The vacuum and gas supply equipment of the multi-pass cell is omitted here for clarity. Double lines represent electric connections, single lines the optical beam path. (DAC - data acquisition card).

The quantitative experiments can be classified as follows: Firstly, trace amounts of CH_4 were measured at elevated pressure (100 mbar) by means of QCLAS and TDLAS (section 4.2.3). Details on the operation conditions for the QCL (Alpes Lasers) emitting at 7.43 μm can be found in table 4.1. Since the QCL had to be temperature stabilised for the different conditions the spectra were not recorded with the same (static) gas mixture. The lead salt laser was tuned in such a way that a spectral overlap of at least one spectral line exists. The analysed spectral features, their line strength S from the *HITRAN* database [80] and the associated experiments are listed in table 4.2. A gas mixture of 1 % CH_4 in Ar was further diluted by Ar to achieve a mixing ratio of ~ 10 ppm.

Next, measurements on a standardised gas mixture of 1117 ppm CO in N_2 using the *intra* pulse mode were performed under low pressure conditions at 0.5 -5 mbar (section 4.2.4). The totally availbale sample volume of this special gas mixture was too small (~ 7 l) compared to the multi-pass cell volume. Therefore an extension of the measurements to elevated pressures were not feasible. The corresponding QCL (IAF Freiburg [81] was operated at -25 °C with a pulse length of 175 ns and a trigger frequency of 1 kHz yielding a spectral scan of 2146.9 - 2147.4 cm^{-1} and covering two lines at 2147.081 cm^{-1} ($^{12}C^{16}O$) and 2147.205 cm^{-1} ($^{12}C^{17}O$) respectively. The spectra were averaged over 50 QCL pulses and compared with TDL spectra covering almost the same spectral range.

A mainly qualitative measurement series where the total pressure was reduced from 100 mbar down to 0.5 mbar was added to link the results from the CH_4 and CO experiments. In this case the dilution of CH_4 in Ar was varied to keep the CH_4 partial pressure and hence the absorption almost constant for the different pressures and to follow the changes in the line shape using the *intra* pulse mode. The corresponding CH_4 number densities ranged from 1.2×10^{15} to 1.8×10^{15} cm^{-3}. In contrast to the earlier CH_4 measurements the QCL heat sink temperature was reduced to -2 °C yielding a higher signal at the detector but loosing the spectral overlap with the lead salt laser which could not be tuned to 1347.0 - 1348.1 cm^{-1}.

Method	Heat Sink [°C]	Pulse Length [ns]	Trigger Freq. [kHz]	Spectral Range [cm^{-1}]	Averages per Spectrum	Time per Spectrum [s]
TEC modul.	6 ... 25	150	5	1345.5 ... 1347.5	1	~ 100
inter pulse	18	16	150	1346.4 ... 1346.9	~ 150	1
intra pulse	15	180	5	1346.1 ... 1347.4	20	~ 0.01

Table 4.1:   QCL operation conditions used for a comparison with a TDL spectrometer. Three different tuning mechanisms were employed to the QCL.

#	Line Position [cm^{-1}]	Line Strength [cm/molecule]	QCLAS	TDLAS
	1346.3301	3.459×10^{-20}	h	h
	1346.5755	2.306×10^{-20}	h[*)	h, l
	1346.7396	3.460×10^{-20}	h	h, l
	1347.0543	5.775×10^{-20}	h	h, l
A	1347.8016	3.381×10^{-22}		
B	1347.9214	3.466×10^{-20}	l	
C	1348.0416	3.470×10^{-20}	l	
D	1348.1525	1.642×10^{-22}	l	

Table 4.2:
CH_4 line positions and line strengths used for the comparison between QCLAS tuning methods and TDLAS measurements [80].

h - experiments at 100 mbar

l - experiments down to low pressure conditions

[*) using the *inter* pulse mode only this line could be analysed

4.2.3   CH₄ detection at elevated pressure

For each of the studied QCL operation modes two spectra were acquired and analysed independently. The mixing ratios were determined from (i) a straightforward integration over the absorption line without any assumption of a distinct line profile and (ii) from a Lorentzian fit to the spectral feature. The results from both approaches and the mixing ratio measured with the lead salt laser are shown in figures 4.3 a - c. Generally, it can be concluded that the mixing ratios obtained from TDLAS and the three QCLAS methods yield a fair agreement at 100 mbar, i.e. at elevated pressure where the line profile is not perturbed by non-linear absorption phenomena. In this case also no significant difference between the two analysis approaches (integration or Lorentzian fit) can be found. It should be noted and will be shown below for lower pressure that both facts are not that obvious as they may appear.

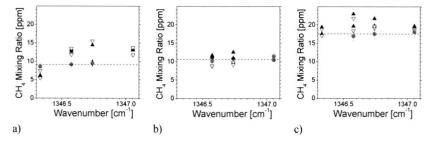

a)   b)   c)

Figure 4.3:   Comparison of CH₄ mixing ratios obtained from TDLAS (circles) and QCLAS (triangles) using different tuning mechanisms (a - TEC modulation, b - *inter* pulse mode, c - *intra* pulse mode) at 100 mbar. The mixing ratios were calculated from the area of a Lorentzian fit to the spectral line (triangle up) or an integration over the absorption feature (triangle down). Averaged TDLAS values (dashed line) serve as a guide to the eye.

However, the results in figures 4.3 a - c also reveal typical uncertainties for each QCL tuning method. The highest difference between TDLAS and QCLAS of about ± 30 % is obtained for the modulation of the TEC controller. The main reason is the non-negligible noise of about 2 % in transmission because of unaveraged spectra. A better agreement is found for the *inter* pulse mode based on the sophisticated *TDL Wintel* software package providing an optimised frequency scan and signal averaging. TDLAS and QCLAS results differ by 5 - 10 %, the scatter between different QCLAS measurements is even less, which is also due to the highest number of average cycles (~ 150) in this assessment. It is important to note that all advantages along with the *inter* pulse method are at the expense of the inherently reduced sweep range (~ 0.5 cm⁻¹). The mixing ratios obtained by employing the *intra* pulse mode differ by 5 - 20 % from the simultaneous TDLAS measurements. The data analysis for this tuning method is very sensitive to the determination of the baseline due to pulse-to-pulse intensity fluctuations. Thus, deviations of 20 % are particularly observed for weaker absorption lines (e.g., at 1346.58 cm⁻¹ and 1346.74 cm⁻¹), where an uncertainty in the baseline causes a bigger error in the integrated absorption coefficient than for the stronger features. In contrast to the previous experiments the CH₄ mixing ratio was about 19 ppm and hence higher than the theoretical value of 10 ppm. The potential reason is a contamination from the rotary pump line. Under static gas conditions this could not be purged, but the absolute mixing ratio is not relevant here. This also demonstrates the advantage of the dual laser setup approach. Assuming that TDLAS yields correct mixing ratios only the deviation of QCLAS results from the counter-measurements is of interest here.

To summarise, at elevated pressure QCLAS yields correct concentrations without additional calibration if line strength values are known. The results are independent from the operation mode which can be chosen depending on whether a high time resolution, an on-line analysis or an increased sweep range is preferred. In some cases, e.g., *in-situ* measurements in atmospheric pressure discharges where broad absorption features are present, the *inter* pulse mode may suffer from its limited sweep range, because the baseline cannot be detected simultaneously next to the absorption line and hence complicates the on-line analysis. Additional baseline acquisition and subtraction is then required. A similar comparison between *inter* and *intra* pulse mode has been conducted in the framework of real-time breath gas detection of NH_3 and C_2H_4 at sample pressures of 200 Torr (266 mbar) or 1013 mbar [59]. The current results are partly supported by the conclusions from this study. Unfortunately, a complementary method was not applied in their case requiring a calibration with standardised gas mixtures. Both calibration and, particularly, the estimation of the effective line width of a pulsed QCL, which is not straightforward (cf. sections 4.3 and 4.4), were not in the focus of the present experiments that aimed at revealing potential differences (e.g. non-linear absorption) between measured and true values.

Measuring at reduced pressure - if possible - provides an increased selectivity and is generally preferable. For this purpose, spectra of the two strong CH_4 absorption features marked with B and C in table 4.2 were recorded while the total pressure was reduced from 100 mbar down to 0.5 mbar. A clear transition from an almost undisturbed line profile at elevated pressure to an asymmetric absorption line at lower pressures can be seen in figure 4.4. The lines exhibit a characteristic oscillatory structure at their low frequency side. In a spectrum which is recorded in the time domain this artefact always appears after the absorption line because of the frequency-down chirped laser. The oscillatory structure, which is also known as fast or rapid passage effect [55, chapter 2], is very pronounced at low pressure conditions and for strong absorption features (e.g., at 0.5 mbar both CH_4 lines with nearly the same line strength exhibit an overshoot in the transmission spectrum of $\sim 40\,\%$ in figure 4.4). With increasing pressure the distortion of the line is damped, but still slightly present at 100 mbar. Moreover, the rapid passage effect which is connected with the fast chirp of pulsed QCLs not only influences the shape of absorption lines, but also their integrated absorption coefficient and hence the deduced molecular number densities. Consequently this leads to an underestimation of number densities and quantitative measurements provide no longer correct absolute densities.

Although the present experiment was chosen to study the appearance of the rapid passage effect qualitatively by means of strong absorptions ($\sim 80\,\%$) it is interesting to compare the calculated mixing ratios from QCLAS and TDLAS experiments. In figure 4.5 their ratio is plotted for lines B and C. This also eliminates the effect of a potentially reduced absorption path length due to the gas flow regime. Similar to the previous experiments at elevated pressure (figure 4.3 c) the QCLAS mixing ratios at 100 mbar almost agree with those from TDLAS. At lower pressures the mixing ratio is considerably underestimated. The discrepancy between QCLAS and TDLAS concentrations increases monotonically from 100 mbar down to the low mbar range. Around 1 mbar only 20 % of the TDLAS value can be determined from the QCL spectra. Similar to these measurements van Helden et al. kept the CH_4 number density constant and varied only the amount of buffer gas and hence the total pressure between a few and 100 Torr (133 mbar) [27]. Below 40 Torr (53 mbar) they also underestimated the CH_4 number density (with a maximum deviation of 16 %) using absorption features at 1253 cm^{-1}.

In figure 4.5 a small deviation of 25 % still remains in the QCLAS/TDLAS comparison at 100 mbar. This might be on the one hand due to the sensitivity of the integrated absorption coefficient on the determination of the baseline yielding a maximum error of ~ 20 %. On the other hand not entirely damped non-linear absorption phenomena (note the still asymmetric line profile) may play a role which is discussed further in the next section.

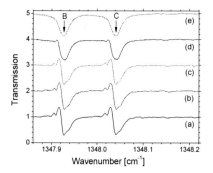

Figure 4.4: Transmission spectra for CH_4 absorption features B and C (table 4.2) measured at (a) 0.5 mbar, (b) 5 mbar, (c) 10 mbar, (d) 50 mbar and (e) 100 mbar employing the *intra* pulse mode. (Spectra are stacked for clarity.)

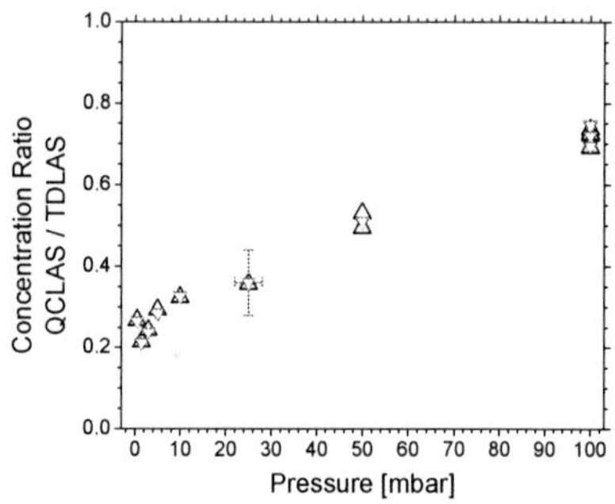

Figure 4.5: Ratio of the simultaneously obtained CH_4 concentrations using QCLAS (*intra* pulse mode) and TDLAS at various pressures. Data analysis was carried out independently for two spectral features: $\triangle$ - 1347.92 cm^{-1}, $\triangledown$ - 1348.04 cm^{-1}.

4.2.4   CO detection at low pressure

It is clear from the previous section that measurements at lower pressure require a correction. Calibration of the line strength or the absorption coefficient are particularly feasible for stable species, but remain difficult for transient molecules. For selected applications (e.g., determination of the gas temperature) the ratio of two lines was found to remain constant for wider pressure ranges [27]. A correction exclusively based on molecular line parameters and the properties of the QCL (e.g., chirp rate) while retaining a calibration-free method would be generally preferable and is essential in metrology [82,83]. Duxbury et al. developed a model taking into account non-linear absorption effects which enabled them to compare experimental and calculated spectra qualitatively [34 and ref. therein]. However, the retrieval of molecular concentrations from an experimentally obtained and disturbed spectrum has not yet been reported. Therefore an alternative approach for correcting underestimated molecular concentrations was evaluated.

The previous experimental conditions, where absorption values exceeded 75 %, were not adequate for a detailed analysis of inaccuracies induced by the rapid passage effect under low pressure conditions. A standardised gas sample of 1117 ppm CO was now studied by means of the dual IR laser arrangement. The pulsed QCL emitting at 4.65 µm [81] was operated in the *intra* pulse mode (figure 4.6) and covered two CO spectral features (figure 4.7). Absorptions between 15 % and 60 % were achieved for the strong line (2147.081 cm^{-1}). For pressures higher than 1.0 mbar the weak CO (2147.205 cm^{-1}) line exceeded the noise level and could also be analysed providing a means to separate beginning saturation effects for the strong line from other obstacles.

Although a standardised gas mixture was used the TDLAS system should provide a backup means to verify the expected CO mixing ratio. Lead salt lasers typically exhibit a smaller spectral width than QCLs (cf. section 4.3). The same absorption line appears typically narrower and stronger for a TDL. Hence, the absorption of the strong CO line was already 35 % at 0.5 mbar total pressure and rapidly approached total attenuation of the radiation whereas the weak CO line could be detected without perturbation under all conditions. The mixing ratios obtained from the saturated CO line measured with TDLAS (figure 4.8) are therefore slightly too low. Additionally, a second weak laser mode of less than 5 % of the total intensity was present. The systematic error induced by this multimode behaviour increases with stronger absorption. Nevertheless, the results from the weak CO line, including ~ 5 % error due to the second mode, confirm the nominal mixing ratio (figure 4.8).

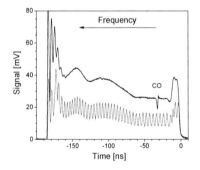

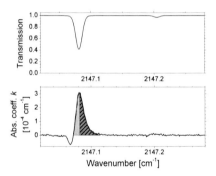

Figure 4.6: Sample CO feature at 2147.08 cm^{-1} acquired in the *intra* pulse mode (1117 ppm, 1 mbar total pressure). The etalon spectrum (FSR = 0.0485 cm^{-1}) provided the relative frequency calibration. Data acquisition was triggered at the end of the pulse, i.e. at its low frequency part.

Figure 4.7: Calculated transmission spectrum (upper) for 1000 ppm CO at 3 mbar (L_{eff} = 24 m) and the corresponding absorption coefficient k measured in the *intra* pulse mode (lower). Both the entire positive area (grey) and the undisturbed high frequency part (hatched) were used for data analysis.

The basic idea to accomplish better results from QCL spectra is to exclude the oscillatory structure during the calculation of molecular number densities. As soon as the absorption line is affected by the rapid passage effect the integral over the absorption line, i.e. the integrated absorption coefficient (eq. (2 - 6)), is reduced and so is the number density and mixing ratio respectively. If half of the absorption line is undisturbed the integration can be carried out only over this part yielding half the actual number density. The maximum of the absorption feature is used as - in fact the only reasonable - indicator to split the measured line into two parts: the unaffected high frequency part is integrated whereas the part including the oscillatory structure is neglected (figure 4.7). The mixing ratios for the CO gas sample obtained by this method are shown in figure 4.9. The results from an entire integration over the positive part of the absorption coefficient k are also displayed. It should be noted that the proposed approach still does not include any calibration since neither the line strength nor the absorption path length L_{eff} are adapted to a reference measurement.

It is obvious from figure 4.9 that integrating the entire disturbed line profile for the strong CO feature underestimates the nominal concentration for all studied pressures: with increasing total pressure from 0.5 - 5 mbar only 85 % to 52 % of the expected mixing ratio is obtained indicating that the influence of the chirped laser pulse increases. If the partial integration is applied, the CO mixing ratio accords with its nominal value up to 1.5 mbar. For slightly higher pressures this approach also fails to give the correct value, but the deviation is reduced compared to a full integration. This behaviour may be explained by a second phenomenon which can occur if (pulsed) QCLs are used. Absorption features are not only affected by rapid passage oscillations caused by the rapidly chirped pulse but also lowered due to power saturation as a consequence of the high laser output power. Consequently, the integration over the oscillation-free half of the line profile gives reduced concentration values.

McCulloch et al. observed power saturation effects in C_2H_4 spectra measured with a pulsed QCL [56]. They used the peak absorbance instead of the integrated absorption coefficient to qualify the non-linear absorption phenomena. In the low pressure regime of a few mTorr (10^{-3} mbar) oscillations of the rapid passage effect were clearly identified. Additionally, a non-linear increase of the absorption with pressure was reported. It was much more pronounced and appeared earlier for transitions with a strong dipole moment or line strength. Moreover they pointed out that with increasing pressure of a buffer gas, firstly, the oscillatory structure is damped, and next, the remaining asymmetric line shape becomes symmetric, but still exhibits power saturation.

This accords well with the experiments using both CH_4 (figure 4.4) and CO (figure 4.9). In both cases relatively strong absorption lines were used (line strength in the 10^{-20} cm/molecule range similar to the strongest feature in [56]). Although the CH_4 absorption line appears almost symmetrically at 100 mbar in figure 4.4 power saturation may explain the remaining discrepancy to the TDLAS measurements. All CH_4 experiments were carried out with a partial pressure of a few tens of mTorr ($10^{14} \ldots 10^{15}$ cm^{-3}) where power saturation was also present for all C_2H_4 lines in [56] including weaker transitions. Since the collisional damping of both rapid passage effect and power saturation is strongly reduced when the total pressure is reduced from 100 to a few mbar the analysis of the CH_4 lines yielded only 20 % of the actual concentration in the previous section (figure 4.5).

In contrast, the CO experiments were carried out with a partial pressure between only 0.4 mTorr (1.4×10^{13} cm^{-3}) and 4 mTorr (1.4×10^{14} cm^{-3}). This appears to be the range for strong absorption lines where power saturation becomes visible, because the (peak) absorption coefficient does not follow linearly the increasing pressure [56]. This was also observed in the present experiments. For total pressures up to 1.5 mbar power saturation can be neglected for the strong CO feature and hence integrating the undisturbed part of the line profile yielded correct CO concentrations (figure 4.9). The entire line profile is only affected by the rapid passage oscillations which can be excluded from the analysis. However with increasing pressure also power saturation should be accounted for and the partial integration is no longer sufficient for a correct analysis. If the weak CO transition ($S = 2 \times 10^{-21}$ cm/molecule) is considered, this pictures changes slightly, because power saturation may occur at partial pressures higher than 10 mTorr (3.3×10^{14} cm^{-3}) which were not studied here. Moreover the rapid passage effect plays a minor role. Consequently, both partial and entire integration over the weak absorption feature approach the nominal concentration at the expense of an increased error since the SNR was 3 under best conditions at 5 mbar total pressure.

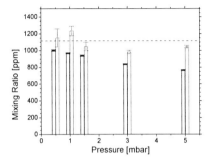

Figure 4.8: TDLAS mixing ratios determined from CO lines at 2147.08 cm^{-1} (black) and 2147.21 cm^{-1} (grey). The dashed line represents the nominal value.	**Figure 4.9:** QCLAS mixing ratios determined from CO lines at 2147.08 cm^{-1} (black) and 2147.21 cm^{-1} (grey) by means of an integration over the entire line (open bars) or only half the line without rapid passage structure (hatched bars). The dashed line represents the nominal value.

If power saturation is present under low pressure conditions, an inhomogeneously (i.e. Doppler) broadened absorption coefficient is reduced by a constant factor (eq. (2 - 32)). This may provide a means to estimate an effective saturation parameter S_{eff} from the CO experiments: at 3 mbar, where the first deviations for the 2147.08 cm^{-1} line are observed (figure 4.9), the Doppler broadening $\Delta \nu_{\text{D}}$ (HWHM, 2.6×10^{-3} cm^{-1}, 78 MHz) still exceeds the collisional broadening by one order of magnitude $\Delta \nu_{\text{L}}$ (HWHM$_{\text{L}}$, 2.4×10^{-4} cm^{-1}, 7 MHz) and the mixing ratio deduced from the partial integration is underestimated by a factor of 0.83. The effective saturation parameter would be $S(\text{CO})_{\text{eff}} = 0.45$.

Using equations (2 - 28) and (2 - 5) the Rabi frequency $\Omega(\text{CO})$ follows from the molecular line parameters and the (peak) power density of the QCL and yields 1.7 MHz. Here an upper limit of the peak power of 500 mW and a spot size of 4 mm diameter was assumed. If the relaxation rates $\gamma_1 \approx \gamma_2 \approx \gamma$ [84] are estimated from the collisional half width (7 MHz) a saturation parameter $S(\text{CO}) = 0.06$ is obtained. Obviously the theoretical value is too low and does not accord with $S(\text{CO})_{\text{eff}}$, but it is still in the same order of magnitude. The discrepancy would increase for 5 mbar, but in this case a purely Doppler broadened line might be insufficient to estimate $S(\text{CO})_{\text{eff}}$ which would become frequency dependent. Nevertheless, the calculated saturation parameter S_{CO} is in the same order of magnitude as $S(\text{CO})_{\text{eff}}$, i.e. the QCL output power is almost as high as required for non-linear absorption phenomena. The high radiation power may lead to self-focusing of the laser [55]. Hence, the power density increases due to a reduced beam cross section and so are the Rabi frequency Ω(CO) and the saturation parameter Σ(CO).

A similar calculation for CH_4 at 0.5 mbar (figure 4.4 a) where the partial integration would underestimate the nominal mixing ratio by a factor of 0.42 gives $\Sigma(CH_4)_{eff} = 4.67$. Assuming the same laser parameters as an upper limit yields a Rabi frequency Ω(CH_4) of 1.5 MHz. The pressure is lower compared to the CO case and thus the relaxation rate reduces to 0.8 MHz. The theoretical saturation parameter ($\Sigma(CH_4) = 3.52$) is now only slightly lower than estimated from the experiment. The nominal mixing ratio would be underestimated by a factor of 0.47 which is almost within the uncertainty of the factor 0.42 extracted from the measurements.

Both examples demonstrate that the (peak) power of pulsed QCLs is typically in a range where power saturation might occur for absorption features with a high transition dipole moment or line strength[3]. This effect is independent from the rapid passage effect due to the chirped laser pulse which was excluded by integrating the undistorted half of the line. It is clear from equation (2 - 27) that transitions with a 10 times weaker line strength or dipole moment yield a 100 times weaker saturation parameter Σ. In this case the absorption follow Beer-Lambert's law. At higher pressures (> 10 mbar) the relaxation rate is increased and the risk of observing non-linear absorption is reduced. A detailed examination of the elevated pressure experiments in the previous section is omitted here, because extracting the frequency dependent saturation parameter in the case of pressure broadened lines would require a thorough line profile analysis. Due to the increased QCL laser line widths (section 4.3.3) this is not straightforward and may cause errors higher than small residual power saturation effects.

It is necessary to discuss the uncertainties for the QCLAS measurements provided in figure 4.9 which appear clearly higher than for the TDLAS experiments (figure 4.8). However the error bars for the QCL results, which were averaged from 4 independently recorded and analysed spectra, represent a maximum uncertainty due to slight changes in the integration limits of the line profile. The apparently large uncertainties in figure 4.9 are mainly caused by the determination of the line maximum as one integration limit. For the weak line (due to the limited SNR) this yields an error of ~ 20 % whereas it is less than 10 % for the strong feature. In contrast, the fluctuations among the 4 different spectra for constant integration limits would be < 1 % for the strong CO line and < 5 % for the weak line. These values can directly be compared with the given uncertainties of the TDL measurements (figure 4.8) which were determined from a plot of the mixing ratio against time.

Generally, it can be concluded that the proposed partial integration over the high frequency part of an absorption line distorted by the rapidly chirped QCL pulse has made possible a calibration-free determination of molecular number densities. This approach however requires that power saturation can be neglected which is particularly the case for transitions with a weak dipole moments. In this case rapid passage oscillations are small and

---

[3] The Rabi frequency and the saturation parameter are defined for the transition dipole moment (eqs. (2 - 27) and (2 - 28)). If transitions of the main isotope with an abundance of e.g. $I_a > 0.95$ are considered, the conclusions are obviously also valid for the line strength (eq. (2 - 24)).

the undisturbed half of the absorption coefficient is not reduced. Although not applicable in all situations, selecting weaker lines and employing a multiple pass cells would be preferable compared to measuring strong lines ($S > 10^{-20}$ cm/molecule for the main isotope) using a short path. These lines would suffer from power saturation for almost all relevant pressures, especially in the field of plasma diagnostics. For weaker transitions non-linear saturation effects exceed the measurement errors at slightly higher number densities than for strong lines which can be already interesting for low pressure applications, e.g., radical detection in low pressure plasmas. As discussed earlier, using weaker absorption features in the *intra* pulse mode increases the uncertainty due to baseline fluctuations and the determination of the line maximum as integration limit.

A more detailed treatment including precise threshold values for the different non-linear effects would be definitely interesting, however, not only the molecular line parameters, but also laser parameters, i.e. output power and chirp rate, influence the observed obstacles. It will be shown in the next section that these laser properties vary for each laser and cannot be adjusted independently. Therefore the experimental conditions, in particular the QCL parameters, have to be carefully chosen to minimise or eliminate non-linear deviations from Beer-Lambert's law for the selected transition.

4.3   Properties of chirped QCL pulses

4.3.1   Chirp rate of pulsed QCLs

As pointed out earlier, the emission frequency of a QCL is determined by the temperature of the device. The tuning behaviour is often characterised by the (negative) temperature tuning coefficient, e.g., given in cm^{-1}/K, which is well suited in those cases where the temperature of the device is known or can be determined easily. In pulsed operation the seed current for the laser generates heat during the nanosecond pulse which in turn makes the frequency decrease throughout the duration of the laser pulse. This frequency-down chirp is present under all conditions and especially exploited in the *intra* pulse mode to gain a spectrum. For the rapid tuning methods of the QCL it is, however, not feasible to measure the temperature accurately and to describe the frequency scanned laser using the temperature as a parameter. The chirp rate dν/dt of the laser is a better means and also a critical parameter to understand the rapid passage effect.

Therefore the chirp rates of 31 pulsed lasers, mainly commercially available and covering almost the entire MIR spectral range, were studied in order to identify and generalise interdependences and limits of working conditions. This would enable an optimised application for spectroscopic purposes. It transpires that the chirp rate is not a constant for a QCL device. Thus, based on the recorded fringes of a Germanium etalon, two different values were defined. Two different lasers may be compared by means of an average chirp rate across a 100 ns interval, $\Delta\nu$/dt_{100}, (figure 4.10). Additionally, the chirp rate dν/dt was determined and followed throughout the entire pulse length (figure 4.11). All lasers were operated clearly above 0 °C, close to its threshold and with the same electronics (Q-MACS).

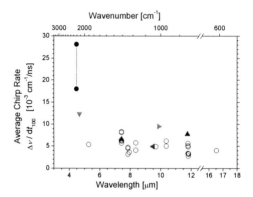

Figure 4.10:   Average chirp rate $\Delta\nu/dt_{100}$ of pulsed QCLs emitting between 600 and 2250 cm^{-1}. Displayed values were averaged over a 100 ns interval. The dashed line represents a range of average chirp rate values achieved by adapting the operation conditions (○, ● - Alpes Lasers, ▼ - IAF Freiburg, ▲ - nanoplus GmbH, ◀ - Laser Components, ▶ - University of Sheffield)

It is obvious from figure 4.10 that no general (average) chirp rate for pulsed QCLs exist. It strongly depends on each device and the thermal behaviour of the active laser zone. Nevertheless, a lower limit of the average chirp rate of about 0.003 ...0.005 cm^{-1}/ns (90 ... 150 MHz/ns) can be estimated from this survey. If appropriate pulses of a few hundred nanoseconds are applied to a pulsed QCL, a corresponding spectrum covering more than 0.5 cm^{-1} can easily be achieved. In contrast, if short pulses (i.e. *inter* pulse method) are applied, the laser frequency is already swept by ~ 0.01 cm^{-1} (300 MHz) during a pulse which additionally serves as a first estimate of the resolution limit of this method. Figure 4.10 also shows an increased chirp rate at lower wavelengths which follows from the behaviour of the Bragg-grating of the QCL. According to the Bragg condition $\lambda_B = 2n_{eff}\Lambda$ the DFB grating with the period Λ selects the emission wavelength λ_B. Temperature tuning of the laser is achieved by a change of the refractive index n_{eff} which shifts the emission wavelength λ_B [65,85]. The corresponding wavenumber shift is then $|\Delta\nu| = \Delta\lambda/\lambda^2$, i.e. the wavenumber shift is wavelength dependent and increases at lower λ as was observed in the experiments.

Further generalisation is not feasible since the chirp rate is not a device constant. As demonstrated for the 4.5 µm laser (figure 4.10), which was operated at two different voltages, the chirp rate covers a certain range and is influenced by several parameters, such as the laser material properties, thermal management provided by the laser design, and the external operation conditions. The pulse parameters, i.e. pulse width, duty cycle and input power, determine the temperature of the laser. The temperature difference between the active zone and the heat sink may range from of a few K for LN cooled heat sinks [86,87] up to tens of K in the case of the room temperature operation [57,73,74,76]. The pulsed laser is in a thermal non-equilibrium state with a rapid increase of the core temperature during the pulse. Characteristic time scales are in the order of a few µs [86,88] which in turn causes a temperature increase and thus a unconstant frequency chirp throughout the laser pulse.

The (unaveraged) chirp rates dν/dt of a few of the studied QCL devices is displayed in figure 4.11 showing no general trend. On the one hand many lasers exhibit a decreasing chirp rate during the pulse, on the other hand QCLs of the same emission wavelength show a diametrically opposed chirp rate behaviour (middle panel). Considering those lasers with a decreasing chirp rate, the relative drop within 150 ns is typically less than 40 % below the initial value. The decrease is often more pronounced at the beginning of the pulse indicating that especially short pulses exhibit a strong chirp rate and that the temperature increase slows down during the pulse.

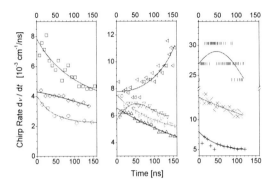

Figure 4.11: Chirp rate dν/dt of different QCLs during a laser pulse (**left**: $\diamond$ - 606 cm^{-1}, $\bigcirc$ - 846 cm^{-1}, $\square$ - 964 cm^{-1}; **middle**: $\triangle$ - 1200 cm^{-1}, $\triangleright$ - 1347 cm^{-1} (Alpes Lasers), $\triangledown$ - 1347 cm^{-1} (nanoplus), $\triangleleft$ - 1347 cm^{-1} (Alpes Lasers); **right:** + - 1900 cm^{-1}, $\times$ - 2140 cm^{-1} [81], | - 2236 cm^{-1}). Note the different vertical scales in the panels.

Notably, different packages were used to mount the sample QCLs chosen in figure 4.11, i.e. the early ST mount ($\diamond$, $\square$, $\triangle$, $\triangleleft$) and the more recent NS mount ($\bigcirc$, $\triangleright$, +, |) provided by Alpes Lasers [89], a TO-8 can ($\triangledown$, nanoplus) or a c-mount ($\times$, IAF Freiburg). A characteristic chirp rate behaviour for a specific package is not observed. The reason may be that the rapid temperature increase in the on-phase of the pulse is mainly determined by the internal design (e.g., cavity dimensions, cladding layers, mounting and material combination) of the laser [73]. The heat sink and package may only support the different measures to increase the heat removal. In contrast, a clear improvement in the QCL thermal management enabling cw room temperature operation in the future can be seen in figure 4.11. One of the early commercially available pulsed QCLs ($\triangleleft$) is the only device which exhibits a continuously increasing chirp rate suggesting a limited heat extraction and self-heating of the device whereas a newer QCL at the same wavelength from the same supplier ($\triangleright$) shows the typical decrease as found for other lasers. For the most recent QCL within this survey (|) the chirp rate varies only by 20 % limited by the scatter of the data. Unfortunately, this is caused by a combination of the strong chirp rate of this 4.5 μm laser and the sampling rate of the oscilloscope.

4.3.2   Adiabatic and linear rapid passage

The measured chirp rates of pulsed QCLs are sufficiently high to observe fast passage along with optical transitions at low pressure conditions [34,90] similar to what is described for magnetic resonances if the external field is rapidly swept over the corresponding transition [91]. Using the lower limit for the (average) QCL chirp rate of 0.004 cm^{-1}/ns (120 MHz/ns) a normalised sweep rate A can be calculated (eq. (2 - 35)): assuming pressures below 10 mbar yields collisional half widths of $< 6 \times 10^{-4}$ cm^{-1}, i.e. the relaxation rates $\gamma_1 \approx \gamma_2 \approx \gamma$ [84] are (1...20) MHz for theses conditions. The normalised sweep rate is then $A = (300 ... 120\,000)$ and thus clearly $>> 1$ being the condition for the appearance of rapid passage effects [91]. Since the lower limit chirp rate was used it can be concluded that rapid passage effects are present with all pulsed QCLs at low pressure conditions.

If the laser intensity is sufficiently high also power saturation, described by the saturation parameter Σ (eq. (2 - 27)), may be observed, i.e. a substantial population transfer occurs between the lower and the upper level of the transition. If the results for CO and CH$_4$ from section 4.2.4 are generalised for strong absorption features ($|\mu| \geq 0.05$ D or $S > 10^{-20}$ cm/molecule for the main isotopes), Rabi frequencies in the order of $\Omega = (0.5...2)$ MHz can be expected for power densities of ≤ 40 kW/m^2 (i.e., < 500 mW peak power for a 4 mm diameter parallel beam). Using the above mentioned relaxation rate γ under low pressure conditions yields saturation parameters of $\Sigma = (0.04 ... 6.3)$. While the lower value is negligible the upper limit case approaches the condition $\Sigma >> 1$ for strong power saturation and may underestimate number densities by 60 %. The situation is even worse if self-focusing of the laser occurs [55].

The combination of strong power saturation with the rapid passage effect this is often referred to as adiabatic rapid passage effect. The corresponding figure of merit is A/Σ [34,91, chapter 2]. As long as $A/\Sigma >> 1$, also known as linear rapid passage regime (table 2.1, iii), the interaction time of the chirped laser with the transition is relatively short. However, if $A/\Sigma << 1$ (adiabatic rapid passage) the interaction time is considerably increased and strong optical pumping occurs. Typical frequency-down chirped QCL pulses are within the linear regime which follows from the obtained estimates for A and Σ and has recently been confirmed [35].

Note, rapid passage and power saturation are separate non-linear phenomena. While rapid passage effects are present for almost all pulsed QCLs at low pressure, power saturation might be absent. Grouiez et al. concluded that no rapid passage could be observed in their experiments with SO$_2$ because they used the average power (duty cycle × peak power) to calculate the saturation parameter [92]. However, the appearance of rapid passage cannot be estimated with the saturation parameter. Power saturation or adiabatic rapid passage was indeed not present, because the line strength was too low ($\sim 10^{-21}$ cm/molecule), but (linear) rapid passage is nevertheless possible for their chirp rate of 15 MHz/ns (0.0005 cm^{-1}/ns). Their SO$_2$ spectra show a weak asymmetry and the residuals exhibit a weak oscillatory structure at the low frequency side indicating that rapid passage is present but suppressed by the limited bandwidth of the detection system (cf. section 4.4).

In practice the discrimination between both effects is relatively difficult, except where the partial integration method is applicable, because the basic theory is based on homogeneously broadened transitions. However at low pressure conditions the inhomogeneous broadening has to be accounted for in the MIR. Duxbury et al. proposed a treatment of the individual velocity components which requires a numerical calculation [55]. Additionally, molecular alignment caused by the linearly polarised laser beam has to be considered [34].

4.3.3   Resolution and bandwidth limits employing pulsed QCLs

Although the chirp rate is often decelerated in the tail of longer pulses the laser does not approach a steady state temperature and hence emission frequency for even longer pulses than demonstrated in section 4.3.1. In complementary experiments the chirp rate $d\nu/dt$ of a 8.3 μm QCL was studied for pulse widths between 30 ns and 1 μs (figure 4.12) and was found to decrease from 0.005 to 0.001 cm^{-1}/ns (150 - 30 MHz/ns). This provides a means to reduce the influence of the rapid passage effect on absorption lines in the *intra* pulse method by tuning the laser such that the absorption feature of interest appears at the end of the pulse. In contrast this may enhance a potential power saturation because the laser stays slightly longer in resonance with the molecular transition.

It is also important to note that the chirp rate not only determines the frequency scan of a laser pulse and the accompanying obstacles, but also influences the resolution of an absorption spectrum. The resolution limit of pulsed spectrometers is a general phenomenon and has been reported much earlier for pulsed diode laser spectroscopy [61]. It is governed by the laser line width. However, it is essential to distinguish between different terms which can be found in the literature: the effective [22,55,57,61] or integrated [4] line width is connected with both the laser chirp and detection system and will be considered below. In contrast, the instantaneous [70] or intrinsic [66,93,94] line width of QCLs is theoretically only determined by the intersubband transition and the design of the QCL. Due to the entirely different operation principle of these unipolar lasers compared to semiconductor diode lasers (e.g., lead salts) the line width enhancement factor α in the modified Schawlow-Townes formula, which describes the spectral width of semiconductor lasers [95-97], is expected to be zero and thus the intrinsic QCL line width is predicted to be ~ 15 kHz (5×10^{-7} cm^{-1}) [93]. Recently, α values slightly different from zero have been reported [98].

First experiments with cw QCLs yielded line widths (FWHM) of ~ 40 MHz (0.0013 cm^{-1}) indirectly determined from a fit to NO lines and using a standard power supply [70]. More recently a 24 MHz (0.0008 cm^{-1}) FWHM spectral width was obtained using the same approach for a room temperature cw QCL [57,99]. Free-running cw QCL line widths of 1 ... 6 MHz (3 ... 20 × 10^{-5} cm^{-1}) were found in combination with low-noise stabilised current sources from heterodyne experiments [94,100,101]. Although they are still the limiting factor, further improvements in current controller technology enabled QCL line widths of 150 kHz (5×10^{-7} cm^{-1}) to be obtained [93] whereas the theoretical value (12 kHz or 4×10^{-7} cm^{-1}) could be confirmed with frequency stabilisation employing a feedback-loop [102].

The effective line width of pulsed QCLs is typically (FWHM) ≤ 1.2 GHz (0.04 cm^{-1}) [1,22,49,53,57-59,66,103-105] but it does not serve as a characteristic constant for a device due to the dependence from the chirp rate $d\nu/dt$. A formula to estimate the effective QCL spectral width and hence the spectral resolution of a pulsed spectrometer is given by McCulloch et al. [106]. However this formula is only valid for particular assumptions and it is therefore interesting to discuss the underlying bandwidth-time product or uncertainty principle that sets the fundamental limit. From the theory of Fourier transforms it is known that the product of the equivalent duration Δt and the equivalent bandwidth Δf ($= \Delta \nu \cdot c$) cannot fall below a constant C

$$\Delta f \cdot \Delta t \geq C. \tag{4-1}$$

The limiting case would be $C = 1$. If standard deviations are used to express the widths, the bandwidth-time product would give $C = (4\pi)^{-1}$ [107]. Relevant cases for spectroscopic purposes are a rectangular time window ($C = 0.886$ [106,108]) and a Gaussian time window ($C = 2\pi^{-1} \cdot \ln 2 \approx 0.441$ [106]) which is also the practical lower limit of the bandwidth-time product. If the uncertainty principle is applied to a QCL pulse, the equivalent duration can be approximated by the pulse duration $\Delta t = t_{on}$. On the one hand a short laser pulse reduces the frequency chirp $\mathrm{d}f/\mathrm{d}t \cdot t_{on}$, on the other hand it increases the equivalent bandwidth. McCulloch et al. define a "best aperture time" to estimate the resolution of a pulsed QCL, which is based on the assumption that the frequency chirp equals the Fourier limited bandwidth

$$\Delta f = \frac{C}{t_{on}} = \left(\frac{\mathrm{d}f}{\mathrm{d}t}\right) \cdot t_{on} = \left(\frac{\mathrm{d}f}{\mathrm{d}t}\right) \cdot \frac{C}{\Delta f}. \tag{4-2}$$

Equation (4 - 2) can readily be transformed into an approximation for Δf [61,106]

$$\Delta f = \left(C \cdot \frac{\mathrm{d}f}{\mathrm{d}t}\right)^{1/2}. \tag{4-3}$$

Consequently, a pulse width t_{on} shorter than the best aperture time cannot increase but may even lead to a worse spectral resolution due to the uncertainty relation. If pulses slightly longer than assumed for equation (4 - 3) are applied, the effective QCL spectral width is also increased since the laser chirp is then the main limiting factor. Both facts were experimentally observed for pulse widths between 3 ns and 15 ns [67] and have to be considered particularly in respect to the *inter* pulse mode.

Using the *intra* pulse mode the spectral width of the entire pulse is not relevant in terms of resolution since the full chirp already contains a spectrum and the spectral resolution within the pulse is of interest. In this case the time resolution of the detection system determines the equivalent duration Δt and may be Gaussian [56]. According to equation (4 - 3) the resolution of such a spectrum is governed by the chirp rate $\mathrm{d}f/\mathrm{d}t$ (or $\mathrm{d}\nu/\mathrm{d}t$) as an external parameter which is independent from the uncertainty principle. The chirp rate of a QCL including its typical decrease in time and the spectral resolution (i.e. effective laser line width) following from equation (4 - 3) is shown in figure 4.12. The limiting cases for a Gaussian time window and $C = 1$ are also plotted.

Although the studied device exhibits a moderate chirp rate (cf. figure 4.10), it is clear that for pulses shorter than 100 ns the resolution cannot be better than 200 ... 360 MHz (0.007 ... 0.012 cm^{-1}) even under the best aperture time assumption. This estimation accords with other publications [22,55,66]. As pointed out earlier, the effective line width is about two orders of magnitude higher than the intrinsic QCL line width observed without frequency stabilisation. Tuning the absorption line to the tail of the QCL pulse increases the spectral resolution to ~ 150 MHz (0.005 cm^{-1}) which still exceeds the resolution of typical diode laser spectrometers (5 ... 100 MHz, 0.0002 ... 0.0033 cm^{-1} [109,110]). These best QCL resolution values are adequate for detecting absorption features at elevated or atmospheric pressures, but at low pressure conditions it is not appropriate and a deconvolution or correction has to be applied. Since at low pressure Doppler broadened absorption profiles typically show spectral widths of less than the given minimum effective line width the pulsed QCL cannot be considered as narrow bandwidth MIR light source. The situation may be even worse because the studied laser was operated close to its threshold, i.e. at higher input power levels the temperature increase in the active zone and hence the chirp rate is higher which in turn degrades the spectral resolution further.

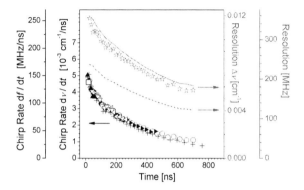

Figure 4.12: Chirp rates of a QCL (black) operated with different pulse widths (□ - 33 ns, ▲ - 85 ns, ▽ - 191 ns, ◁ - 292 ns, ▶ - 492 ns, ○ - 794 ns, + - 997 ns) and the corresponding resolution in a spectrum (grey). The bandwidth-time uncertainty (eq. 4 - 1) determines the achievable resolution and is displayed for three cases: the upper limit ($C = 1$, solid line), the lower limit assuming a Gaussian time window ($C = 0.441$, dashed line) and the intermediate case for a rectangular time window ($C = 0.886$, ☆).

4.4 Bandwidth effects using short QCL pulses

4.4.1 Overview of measurement systems in the literature

A straightforward transfer of the conventional current ramp tuning of semiconductor lasers, among them tuneable lead salt lasers, to QCL spectrometers has lead to the development of the *inter* pulse method using short laser pulses. It was the preferred approach during the first spectroscopic application of a QCL combined with wavelength modulation spectroscopy [1] and enables a time resolution of milliseconds using a sophisticated sweep integration method [22]. Employing this method is not only of interest for on-line trace gas measurements, on which most of the experiments carried out so far were focussing [22,23,49,50,53,57-59,67,71,72,103-105,111,112], but also for monitoring species in plasma diagnostics [25,28,29]. Due to the typically high threshold currents at room temperature (up to a few A) and the considerably higher compliance voltages of QCLs (> 6 V) compared to TDLs the application of short current pulses reduces the input power deposited in the active zone of the QCL and hence the thermal stress of the laser. In contrast to cw tuneable lead salt lasers where the driving current is ramped above threshold, the cw current ramp impressed on pulsed QCLs is clearly below their threshold ($\sim$ mA), but sufficient to tune the laser temperature and frequency. Three different time scales have to be considered in the *inter* pulse mode: the applied power is mainly injected into the active region in a short pulse of $\sim$ 10 ns, cooling occurs during the subsequent μs (cf. section 4.3.1) whereas the heat sink temperature is varied on the ms scale by the continuous current through the laser.

Apart from pulse-to-pulse fluctuations inherent to the pulsed operation spectral line broadening was identified as a main challenge while using the *inter* pulse mode for highly sensitive detection schemes [23]. Effective laser line widths (FWHM) between 0.0095 cm^{-1} (290 MHz) and 0.093 cm^{-1} (2.79 GHz) have been reported and are summarised in table 4.4 (section 4.5). In some cases an asymmetric spectral laser output was deduced [50,72]. According to the discussion of the previous section the resolution of short pulse spectrometers is either Fourier limited for extremely short pulses (5 ns, 290 MHz [49]) or limited by the frequency chirp of the laser (70 ns, 2.79 GHz [50]).

A detailed analysis of particularly laser line width effects in the *inter* pulse mode was carried out by Nelson et al. [22]. Their approach was twofold, i) empirically and ii) theoretically in nature. Theoretical modelling encompassed all above mentioned relevant time scales and predicted an effective laser line width (FWHM) of 0.133 cm^{-1} (4.0 GHz) for a 10 ns pulse exhibiting an estimated chirp rate of 0.019 cm^{-1}/ns. A comparison with measured spectra yielded typically half the predicted values. Firstly, this may be due to a slightly overestimated chirp rate for their 1900 cm^{-1} device operated close to the threshold (cf. figure 4.10 and 4.11). Secondly, extracting of accurate laser line widths from their spectra was difficult, because a strong asymmetry of absorption lines was found. This effects was amplified by applying higher operation voltages, although the simultaneously recorded electrical pulse shapes showed no significant difference. Consequently, Nelson et al. concluded that a QCL should be operated close to its threshold with short pulses close to the uncertainty broadening limit with rise times faster than 8 ns achieved with the original Alpes Lasers operation unit. A minimum laser line width (FWHM) of 0.014 cm^{-1} (420 MHz) was predicted and accords with the results of figure 4.12. The investigation of asymmetrically broadened absorption lines was continued by McManus et al. since such absorption features revealed another phenomenon: the peak absorption signal decreased as the distortion became more pronounced [57]. In these experiments the operation voltage was varied at constant pulse width. McManus et al. suggested the product of peak absorbance and detector signal as a figure of merit for determining the optimum operation conditions. Since the latter factor increases with higher QCL voltages this product levels off which was chosen a the operation point.

Nevertheless, several questions remained unclear from these studies. Firstly, although Nelson et al. worked at low pressure (4 Torr) no rapid passage signatures were found in the spectra. Secondly, the asymmetric absorption lines exhibited both a high frequency tail [22,53] and a low frequency tail [57], even though only the latter case might be explained by the frequency-down chirp of the QCL. A potential pulse-to-pulse frequency jitter or not yet identified properties of the individual pulse were proposed [22].

Recently, the asymmetrical line shapes have been attributed to the detector bandwidth [54]. The experiments presented here focussed on both a detailed analysis of each short chirped pulse and bandwidth effects. Note that for all results of the previous sections the bandwidth of the employed fast detection and digitising system (1 GHz) was assumed to be sufficient, i.e. the instrumental broadening was smaller than the observed effective laser line widths. Different situations are examined in what follows. The comparison of QCL tuning methods including the *inter* pulse mode using pressure broadened absorption features in section 4.2.3 revealed no significant difference to the expected concentrations. Now experiments employing short QCL pulses in the low pressure range are discussed resulting in an entirely different picture.

4.4.2   Experimental approach

The optical setup and dual laser arrangement (QCL and TDL) was essentially the same as described in section 4.2.2. Current ramp, QCL pulses and the gate-off period were synchronised by the *TDL Wintel* software package. In contrast to Nelson and McManus et al. the laser was operated with a Q-MAC system. The pulsed current through the laser is not directly accessible with this QCL driver. Complementary measurements with a miniaturised inductive coil directly at the laser yielded a rise time of the pulse unit of ~ 2 ns. Nelson's idea to reduce the laser chirp with shorter pulses is thus realised since this rise time is faster than for the Alpes Lasers Starter Kit for which a value of 8 ns was confirmed also with our test equipment. Although the QCL was the same as in the previous experiments the emission range was shifted to 1347.75 ... 1348.25 cm^{-1} covering 4 CH$_4$ absorption features which are marked as A - D in table 4.2. Calculated spectra for two series of experiments are shown in figure 4.13.

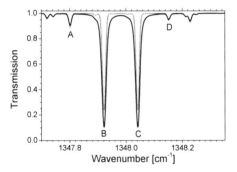

Figure 4.13:   Calculated spectra for CH$_4$ corresponding to the experimental conditions and assuming an instrumental broadening of (FWHM) 0.01 cm^{-1} (~ lower limit): i) 100 % CH$_4$, p_{total} = 10 mbar, L_{eff} = 0.15 m (black); ii) 14 % CH$_4$, p_{total} = 0.5 mbar, L_{eff} = 24 m (grey). Absorption lines are labelled according to table 4.2.

At first, the characteristics of the tuning method in combination with the standard configuration (i.e. detector signal measured with DAC and displayed with *TDL Wintel*) was established by means of a reference cell filled with CH$_4$ at 10 mbar. A rise time of ~ 80 ns (~ 5 MHz bandwidth) was experimentally confirmed for the DAC (National Instruments, PCI-6110E, [113]). The TE cooled detector (VIGO, PDI-2TE-10.6) was used with a custom-made preamplifier (250 MHz). The influence of both QCL pulse width and QCL voltage on the recorded spectra was analysed.

Next, the detection system was extended in order to scrutinise each QCL pulse. A fast and high bandwidth digitising oscilloscope (LeCroy WR104Xi, 1 GHz) recorded all optical pulses of a laser sweep with a time resolution of 0.1 ns. A post-measurement analysis yielded the width of the emitted pulse, the detector signal at constant temporal positions during the chirped pulse as well as the time-integrated signal. Different detectors which are summarised in table 4.3 were employed to check potential bandwidth effects on the results. In contrast to the first experiments using the DAC, the combination of detector and preamplifier is now limiting the bandwidth which was varied between 5 MHz (= DAC value) and 600 MHz. All measurements were carried out under flowing gas conditions at 0.5 mbar total pressure in a multi-pass cell (figure 4.2) aligned to 24 m absorption length. For both the reference and multi-pass cell experiments rapid passage oscillations should therefore be present in the spectra. The studies of the *inter* pulse mode were again backed by complementary TDLAS measurements at the strong CH_4 features around 1346 cm^{-1} (table 4.1).

Detector Type		Cooling	Active Area [mm^2]	Bandwidth [MHz] (Detector & Preamp.)
VIGO	PDI-2TE-10.6	TE	0.25×0.25	250
VIGO	PDI-3TE-10/12	TE	1×1	600
Kolmar	KV104-0.1-A-3	LN	0.1×0.1	≤ 50
Judson	J15D22-M204-S01M-60	LN	1×1	5

Table 4.3:   Detectors and their main parameters employed for *inter* pulse experiments. The first device was used for the reference cell measurements (section 4.4.2 and 4.4.4).

4.4.3   Concentration measurements using the conventional *inter* pulse technique

QCL spectra of the reference cell are shown in figure 4.14 for different operation voltages up to 0.6 V above threshold and a constant pulse width of 20 ns. Since the heat sink temperature was not corrected for the increased input power at higher voltages the spectrum is slightly shifted which can be detected from the spectral position of the gate-off period (figure 4.14). Instead of the expected typical oscillatory structure of the rapid passage effect another distortion of the absorption lines is found in the spectra. A strong asymmetry at the low frequency side of the line, particularly at voltages clearly above threshold, is observed and hampers a reasonable fit to line profiles for a quantitative analysis. The absorption lines are broadened by features similar to a shadow of the original line. This is clearly visible at 0.4 V above threshold for line C, i.e. at the end of the current ramp, and at slightly higher operation voltages for line B. Additionally, the peak absorption decreases as the distortion becomes more pronounced. Only at operation voltages close to the threshold of the QCL (trace a) in figure 4.14) the absorption line exhibits a behaviour similar to the calculation (figure 4.13), i.e. a symmetric line shape and > 50 % absorption, but at the expense of a considerably reduced detector signal. This in turn leads to a poor SNR, even without additional losses from multi-pass optics, and the features A and D vanish below the threshold. Similar obstacles have already been reported (cf. previous section). Since another QCL driver was used here it can be assumed that the observed effects are not directly connected to the current pulses (shape, rise time, etc.).

The acquired spectra were quantitatively analysed by applying two independent methods. Firstly, *TDL Wintel* calculated the CH_4 mixing ratio from a least-square fit of the absorption lines to Voigt profiles simulated by using first principles. These pre-calculated spectra are based on pressure, temperature, effective laser line width and line strengths from the *HITRAN* database [80] as input parameters. Secondly, the mixing ratios were deduced from an additional calculation of the integrated absorption coefficient. In the latter case no line profile or effective laser line width has to be assumed and even the distorted features are entirely captured. Both results are displayed for CH_4 lines B - D in figure 4.15 where no significant difference between the approaches is observed. Results from the fit yielded typically slightly lower values than for the integration. The difference of not more than 10 % may be due to an independent determination of the baseline and slightly different effects of the distortion on the mixing ratios. At higher QCL voltages the integration method is better suited to account for the distortion. Nevertheless, considering the strong absorption features B and C leads to an underestimation of the mixing ratio by at least a factor of 3 or more. The expected concentration is only obtained for the weak feature D at QCL voltages close to the threshold. If the operation voltage is increased the deduced mixing ratio from D also drops to less than 40 % for both methods. Note that the fit results are not considerably changed if the effective laser line width is adjusted in a range which yields a reasonable fit to the acquired spectra (i.e., 0.012 ... 0.020 cm^{-1}, FWHM). Those effective spectral widths are higher than estimated from figure 4.12 for very short QCL pulses.

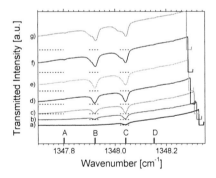
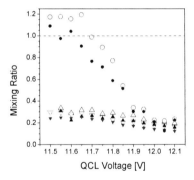

Figure 4.14: *Inter* pulse mode spectra of a reference cell filled with 10 mbar CH_4 measured at constant pulse width (20 ns) of the QCL current and variable QCL voltages: a) 11.5 V, b) 11.6 V, c) 11.7 V, d) 11.8 V, e) 11.9 V, f) 12.0 V, g) 12.1 V. Spectra are stacked for clarity and the corresponding zero signal level from the gate-off period is indicated (dotted lines). Absorption features are marked with A - D (table 4.2).

Figure 4.15: CH_4 mixing ratios deduced for the absorption features B (▲, △), C (▼, ▽) and D (●, ○) from figure 4.14 applying QCL voltages between threshold and 12.1 V. Results were obtained from a fit to a Voigt profile (full symbols) and from an integration over the absorption line (open symbols). Dashed line represents the reference value.

It is clear from the quantitative analysis in figure 4.15 that even for weak absorption features, where power saturation effects can be neglected in a first approximation, the integrated absorption coefficient and hence the concentration values are not conserved, but underestimated for relevant operation conditions and output power values of the QCL. The analysis approach is not critical in this case. Moreover, the picture does not change if the current pulse width is reduced from 20 ns down to 12 ns (minimum) while keeping the QCL voltage near the simultaneously changing threshold. This also indicates that the chirp-limited resolution can be only a part of the explanation.

4.4.4   Consequences of the frequency chirp

All previous experiments and their analysis suggested that a deeper insight into the intrinsic properties of the laser pulses and their detection are necessary to describe the observed obstacles. For this purpose the laser sweep consisting of 230 pulses in the gate-on and 20 pulses in the gate-off phase was monitored with high time resolution. Each QCL pulse represents a spectral data point (or channel) and is described by a time scale t^{Ch} which is connected with the corresponding trigger (figure 4.16). Another time scale t_{pulse} is introduced in figure 4.16 for further data analysis to eliminate the delay between trigger and laser pulse caused by the electronics and the folded optical beam path. Figure 4.17 shows typical examples of several QCL pulses recorded with the CH_4 cell (left panel) or a Germanium etalon (FSR = 0.048 cm^{-1}, right panel) in place. The spectra of the chirped pulses were recorded with similar operation conditions as in figure 4.14 c (i.e. voltage 0.2 V above threshold, 20 ns current pulses), where a reasonable output power of the QCL could be achieved and first line shape distortions were observed. In what follows the diagrams are based on these experimental conditions which were selected as an example, though the discussion and conclusions can be generalised[4]. The channels in the left panel of figure figure 4.17 display the highly time resolved detector signal around absorption feature B. In the right panel a modulation of the QCL pulses by the etalon fringes is visible indicating that the chirp even for these short pulses is high enough to encompass up to a full FSR (figure 4.17 o).

Next, the detector signals of all individual QCL pulses were analysed at a constant time t_{pulse} (slightly averaged over 0.3 ns) and plotted against the channel number (figures 4.18 and 4.19). In figure 4.18 the resulting spectra are shown for the CH_4 reference cell for different t_{pulse} covering the full laser pulse width. Additionally, the entire detector signal was averaged over the pulse width (trace g). The compilation of these spectra reveals several key features of this tuning method:

i)   The rapid passage effect is present in the spectra.
ii)   The time-averaged signal exhibits the same distorted line shapes without rapid passage oscillations as observed with the conventional approach (figure 4.14).
iii)   The peak absorption decreases at higher t_{pulse} and in the averaged signal.
iv)   The position of spectral features (i.e. channel number) shifts with t_{pulse}.

---

[4]   A set of diagrams for 11.5 V QCL voltage (at threshold) can be found in the appendix (figures B.01-B.06).

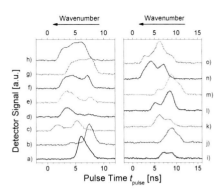

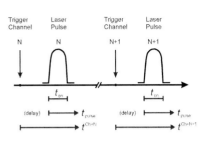

Figure 4.16: Schematic diagram of two QCL pulses (N and $N+1$) of length t_{on} during a laser sweep using the *inter* pulse mode. Two different time scales, i.e. a time scale t^{Ch} for each channel and an intrinsic time t_{pulse} for the laser pulse, are introduced for further data analysis.

Figure 4.17: Time resolved spectra of single frequency-down chirped pulses during an *inter* pulse laser sweep. **Left:** reference cell (0.15 m, 10 mbar CH_4); **Right:** Ge etalon (FSR = 0.048 cm^{-1}). The corresponding channel numbers are a) 95, b) 98, c) 101, d) 104, e) 107, f) 110, g) 113, h) 116 and i) 40, j) 70, k) 100, l) 130, m) 160, n) 190, o) 220 (cf. fig. 4.18/4.19).

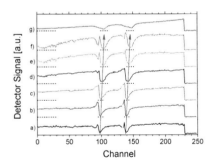

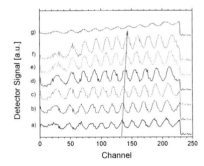

Figure 4.18: CH$_4$ absorption spectra (10 mbar, 0.15 m) determined from the detector signal at t_{pulse} = const. during a chirped QCL pulse: a) 1.0 ns, b) 1.5 ns, c) 2.0 ns, d) 2.5 ns, e) 3.0 ns, f) 3.5 ns, g) averaged signal over the entire pulse. The zero level from the gate-off period is indicated by dotted lines. Arrows represent the shift of spectral features with t_{pulse}. (U_{QCL} = 11.7 V).

Figure 4.19: Etalon spectra (FSR = 0.048 cm^{-1}) determined from the detector signal at t_{pulse} = const. during a chirped QCL pulse: a) 1.0 ns, b) 1.5 ns, c) 2.0 ns, d) 2.5 ns, e) 3.0 ns, f) 3.5 ns, g) averaged signal over the entire pulse. The shift of the extrema with t_{pulse} is marked with an arrow. (U_{QCL} = 11.7 V).

Although not seen in the spectra acquired with the conventional *TDL Wintel* and DAC equipment the typical rapid passage structure at the low frequency side of a line, which is expected under low pressure conditions with pulsed QCLs, is present (i). Moreover, the amplitude of the oscillations decreases as t_{pulse} approaches the end of the short pulse, because the chirp rate dν/dt drops considerably during the first nanoseconds of a pulse[5]. The variable chirp rate is accompanied by an increase of the output power of the QCL at higher t_{pulse} which may cause additional power saturation, at least for the strong features B and C. Consequently the peak absorption decreases at the end of the short pulse (iii). The shift of spectral features (iv) is linked with the chirp of the pulse and the sign of the current ramp slope and will be discussed below. It is nevertheless obvious from figure 4.18 that contributions to the spectrum from high t_{pulse} values are overestimated in the time-averaged detector signal (trace g) due the higher output power at this time. Consequently, the averaged absorption profile appears asymmetrically broadened with a shadow of the line at lower channels. If the temporal development of the laser intensity is reversed, i.e. decreasing intensity during t_{pulse}, which was observed for another QCL, the shadow is observed at higher channels [see also 22,53]. It is also clear that this artificial asymmetric instrumental broadening is difficult to describe analytically and reduces the peak absorption in addition to potential power saturation effects. Thus, concentration values are further underestimated and weak absorption features are dropped below the noise level (e.g. features A and D in figure 4.18 g and 4.14). Averaging also eliminates rapid passage structures in the spectra (ii) and considerably degrades the quality of etalon spectra (figure 4.19). All traces for a constant t_{pulse} in figure 4.19 show a deep modulation of the laser intensity whereas the averaged signal (trace g) is blurred due to the shift of the extrema.

It transpires from the current analysis that the observed line broadening and obstacles are caused by neglecting the intrinsic properties of the laser pulse rather than pulse-to-pulse fluctuations as alternatively proposed by Nelson et al. [22]. Particularly, the laser chirp combined with the current ramp temperature tuning induces complex dependences. In order to examine a few aspects, firstly, the shift of a constant spectral position ν_0 within the ramp was quantified and, secondly, the pulse width of the individual pulses was determined. For this purpose, the channel number of the CH_4 lines B and C and of selected fringes in figures 4.18 and 4.19 were plotted for several t_{pulse} = const. in figure 4.20. It is found that a selected ν_0 = const. shifts by at least 8 ... 10 channels if the frequency chirp of the short pulse is deconvoluted as in figure 4.18. This corresponds to a shift of ~ 0.025 cm^{-1} (750 MHz) for the present experimental conditions and accords well with the effective line widths discussed in the introduction of this section.

Surprisingly, the spectral positions in figure 4.20 shift to higher channel numbers which are typically connected with higher laser frequencies, because the DC laser current is ramped down which increases the frequency. However, the frequency chirp of the short pulse, which induces in fact the shift, leads to lower frequencies.

Next, the start and the end of all individual QCL pulses in a sweep were determined on the time scale t^{Ch} (figure 4.21, right hand scale). A detector signal just above noise level was chosen as criterion for determining both values. The difference yielded the pulse width t_{on} which is clearly different from the nominal current pulse width of 20 ns (figure 4.21, left hand scale). Although the scatter in the t^{Ch} data of ± 2 ns in figure 4.21 is relatively pronounced, the optical pulse width (being actually the difference of the scattered start and end point data) yields only a small uncertainty of ± 0.6 ns. This indicates that mainly jitter between the trigger

---

[5]   Unfortunately, this behaviour is partly clipped in figure 4.18 due to the bandwidth limit of the used detector preamplifier (section 4.4.5), but it was observed in several other experiments which are not displayed here.

event and the laser emission causes the strong fluctuations in t^{Ch} while the pulse width is relatively stable. The detailed analysis shows that the optical pulses appear earlier and fade away later. Consequently the pulse width rises from 4 ns to 10 ns in figure 4.21. This behaviour emanates from the temperature tuning via the DC current ramp while the QCL operation voltage and hence the pulsed QCL current is kept constant. According to figure 4.1 b the temperature is reduced during the ramp which gradually lowers the threshold. If the same pulsed current is fed to the laser, the QCL starts emitting earlier and longer as the channel number increases. It is, however, below the threshold during at least half of the nominal current pulse width for the present conditions. This may lead to the conclusion that the current pulse width can be chosen arbitrarily as long as the operation voltage is kept low and optical pulse remains short enough. In complementary measurements of the pulsed laser current it was found that the current pulse width also plays an important role since it determines the amount of power deposited inside the QCL and hence its heating and chirp rate which in turn influences the short optical pulse. Therefore the current pulse width should be set properly.

Taking into account a lower limit of ~ 0.004 cm^{-1}/ns for the chirp rate an intrinsic and unconstant chirp of the pulses between 0.02 and 0.05 cm^{-1} is found which is even for these short pulses bigger than typical Doppler broadened absorption features. Figure 4.17 demonstrates this for CH$_4$ feature B which is present in more than 20 channels. Notably, all graphs were recorded only 0.2 V above the threshold where the output power and hence the detector signal were just sufficient for achieving a reasonable SNR and thus the described obstacles can hardly be reduced. The examples in the appendix B (laser at threshold, pulse width 0 ... 6 ns, figs. B.01-B.04) confirm that even in this case the laser chirps between 0 and 0.03 cm^{-1}.

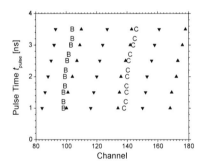
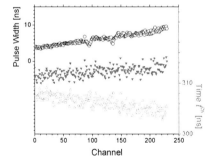

Figure 4.20: Shift of the position of maxima (▲) and minima (▼) of the deconvoluted etalon spectra (figure 4.19) and of the strong CH$_4$ lines (B, C) in figure 4.18. ($U_{QCL} = 11.7$ V).

Figure 4.21: Start (△) and end (▼) point of the QCL pulse on the time scale t^{Ch} (right hand scale, grey) and the corresponding difference yielding the pulse width (○, left hand scale, black). ($U_{QCL} = 11.7$ V).

In conclusion, it is clear from the discussion in this section that a generalised theoretical description is almost impossible to obtain since many parameters are strongly dependent on the single QCL device. Additionally, complex dependences exist between the operation conditions (e.g., current pulse width, DC current ramp, etc.) and the subsequently achieved chirp rate and optical pulse width. The resulting total chirp in combination with the DC ramp parameters finally induce a shift of each spectral component over several detection channels. The total shift corresponds with what is normally considered as effective line width of pulsed QCLs (≤ 0.05 cm^{-1} or 1.5 GHz, FWHM). It was found that spectra recorded on-line with the conventional *TDL Wintel* and DAC equipment were very similar to those which were deduced by averaging the entire detector signal of the individual chirped pulses in the post-measurement analysis. The shift of spectral components appears then as spectral broadening of absorption lines. Since the shape of the optical pulses never approaches a top-hat structure, i.e. the output power is not constant during the short pulse, such an averaging of detector signals leads to an asymmetrical instrumental broadening where the resulting shadow of an absorption line may be observed on both its low or high frequency side. The spectral broadening is also accompanied by a reduced peak absorption. In lieu of an asymmetric instrumental width the sum of all phenomena may be considered as quasi-multimode behaviour since the laser exhibits indeed no single mode emission due to its chirp.

4.4.5   Bandwidth effects

The similarity between the conventionally acquired and artificially averaged spectra (figure 4.14 c and 4.18 g) indicates that the entire experimental setup based on *TDL Wintel* encompasses an unexpected time-averaging of the signals or is limited in its bandwidth. This question is studied in what follows. For this purpose, the same CH$_4$ features (A - D) were measured under flowing gas conditions in the multiple pass cell using three different detector-preamplifier combinations (table 4.3). In all cases these combinations served as the bandwidth bottleneck exhibiting values of 600, 50 and 5 MHz. For high bandwidth detectors the performance of the oscilloscope was degraded down to 200 and 20 MHz respectively. The same data analysis (i.e. deconvolution) as in the previous section was carried out. Since the best indicators to assess the consequences of bandwidth effects are the transients of the rapid structure, the detector signal at $t_{pulse} = 1.5$ ns, where the oscillations are most pronounced due to the high chirp rate, was chosen for a comparison (figure 4.22). Additionally, the time-integrated detector signals and the transmitted intensity measured with the *TDL Wintel* and DAC based system is displayed in figure 4.22.

Firstly, the spectra acquired with the 600 MHz device consisting of a fast TE cooled detector (VIGO, PDI-3TE-10/12) and an appropriate preamplifier (figure 4.22, left panel) is considered. If the oscilloscope bandwidth is reduced (traces e → d → c) the rapid passage peaks are first damped and finally, at 20 MHz, completely suppressed. This bandwidth value already causes an effective averaging yielding similar spectra and asymmetric line shapes as obtained with the *TDL Wintel* system (5 MHz, trace a) or the time-integrated detector signal (trace b). The critical or cut-off bandwidth should be thus in the order of 200 MHz. The effective averaging in the low bandwidth cases also reduces the peak absorption and hence the sensitivity.

The middle panel of figure 4.22 shows results from a 250 MHz preamplifier, which was essentially the same as in the static reference cell experiments of the previous section, and a LN cooled photodiode (Kolmar). Although the detector area was already considerably reduced, the detector elements still limited the total rise time or bandwidth. Consequently, the change of the oscilloscope bandwidth limit (1 GHz → 200 MHz, traces j → i) does not affect the corresponding spectra. A comparison between the 20 MHz, time-integrated and DAC spectra (traces h, g, f) leads to the same conclusion found in the 600 MHz case.

In the right panel of figure 4.22 the spectra obtained with a combination of a LN cooled detector and preamplifier, which are typically used with lead salts (i.e. relatively big detector area and hence slow response time, 5 MHz preamplifier), are plotted. Due to the slow response time the maximum of the ~ 60 ns long detector signal was used for the analysis (trace m). No difference between all three spectra (maximum, time-integrated signal, DAC) is found indicating that in this case the detector already causes distorted line shapes because its slow response time determines actually the averaging process.

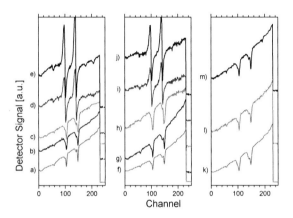

Figure 4.22:   CH$_4$ spectra (0.5 mbar, mixing ratio = 0.14, 24 m) measured by using a detector/preamplifier with an effective bandwidth of 600 MHz (left), 50 MHz (middle) and 5 MHz (right). The oscilloscope bandwidth was 20 MHz (c, h, l), 200 MHz (d, i) and 1 GHz for all other traces. Traces (c, d, e) and (h, i, j) represent the detector signal at t_{pulse} = 1.5 ns whereas (b, g) show the time-integrated signal; (l, m) were taken at t_{pulse} = 22 ns (~ maximum) due to the slow rise time of the detector. Additionally, the detector signals were measured with the *TDL Wintel* system (a, f, k). (U_{QCL} = 11.7 V)[6].

It is clear that the conventional *TDL Wintel* based TDLAS systems and their DACs perfectly match the bandwidth of the detectors employed with this class of semiconductor lasers. However, the straightforward replacement of laser and detector is not sufficient to avoid the observed obstacles in QCLAS. Even if detectors with appropriate rise times are used, which is usually not the case (see discussion at the end of this section), the digitising system has to be adapted as well. The total bandwidth of the detection system consisting of detector, preamplifier and digitiser is the relevant parameter. An estimate for the required minimum bandwidth of the entire system can be derived by means of equation (4 - 1). By analogy with the calculation of the spectrometer resolution earlier in this chapter the frequency chirp $\Delta f (= \Delta v \cdot c)$ of the laser is estimated by its chirp rate and pulse width. The

---

[6]   The same analysis for measurements at the threshold can be found in the appendix B (figure B.05).

time window Δt is now approximated by twice the rise time t_{rise} of the digitiser which is determined by its bandwidth (BW) [114]

$$\Delta f = \frac{df}{dt} \cdot t_{on} ; \quad \Delta t = 2 \cdot t_{rise} = 2 \cdot \frac{0.35}{BW} . \qquad (4 - 4)$$

For simplicity it is also assumed that $t_{on} = 2 \cdot t_{rise}$. Using equation (4 - 4) the uncertainty relation (4 - 1) can be transformed into

$$C \approx \left(\frac{df}{dt} \cdot \left(2 \cdot \frac{0.35}{BW} \right)^2 \right) \qquad (4 - 5)$$

which yields now a chirp rate dependent minimum bandwidth BW_{min}

$$BW_{min} \approx 0.7 \cdot \left(C^{-1} \frac{df}{dt} \right)^{1/2} . \qquad (4 - 6)$$

For a moderate chirp rate of $0.005 \text{ cm}^{-1}/\text{ns}$ (150 MHz/ns) and $C = 1$ a minimum total bandwidth of 270 MHz is required for a spectrometer. This value accords with the cut-off bandwidth estimated from figure 4.22. If the bandwidth is reduced by a single component of the entire detection system, absorption features are asymmetrically broadened for all pressures due to the laser chirp. At low pressure conditions rapid passage structures are additionally clipped.

The absorption spectra obtained with different detectors (figure 4.22) were also quantitatively analysed. The results are compared with simultaneous concentration measurements of a TDLAS system in figure 4.23. For this purpose a weak (D) and a strong CH_4 line (C) were chosen and analysed using the integrated absorption coefficient K. The promising approach of a partial integration over the undisturbed part of an absorption feature was not applied here since many of the deconvoluted spectra in figure 4.22 are unaveraged and hence rather noisy in contrast to traces measured with *TDL Wintel*. A partial integration would therefore increase the relative error of the results. The mixing ratios deduced from spectra using the detector signal at $t_{pulse} = 1.5$ ns or the time-integrated signal and the concentration from a fit of the *TDL Wintel* based system are shown for each detector in figure 4.23.

It is obvious that the mixing ratio of about 14 % CH_4 is substantially underestimated by almost one order of magnitude if the strong line C is used for the analysis (figure 4.23, left panel). At a QCL voltage, which is only 0.2 V above the threshold, no significant difference is found between the results from the different detector signals (time-averaged or at 1.5 ns). The *TDL Wintel* fit yields even smaller concentrations because the fitted line profile covers only a part of the distorted line shape. For measurements at the threshold (figure B.06, left panel) the mixing ratios at 1.5 ns are slightly higher than time-averaged, but still nearly a factor of 10 underestimated. Similar results were obtained for strong absorption features in the *intra* pulse mode under low pressure conditions (section 4.2.4) which clearly indicates that not only the rapid passage effect, but also power saturation plays an important role in addition to bandwidth effects. Even at the threshold the influence of these effects cannot considerably be reduced and a calibration would always be necessary.

Consequently, a better agreement between TDLAS and QCLAS is found for the weak CH_4 line D (figure 4.23 and B.06, right panel) where power saturation is negligible. Particularly, the fit results obtained from *TDL Wintel* at the threshold yield a fair agreement with the TDLAS results as long as the effective laser line width is adapted as a fit parameter (figure B.06). Due to a minimised laser chirp at these operation conditions the absorption feature could be approximated by a symmetric line profile with an increased instrumental broadening which was typically set to (FWHM) 0.012 cm^{-1} (360 MHz). At higher QCL voltages the fit results already fall short of the TDLAS values by at least 25 % (figure 4.23). Mixing ratios calculated from the time-integrated detector signals are found to be lower than determined at $t_{pulse} = 1.5$ ns, but higher than deduced from the fit procedure suggesting that the area of the absorption features is not conserved if the bandwidth of the detection system is limited inadequately. Unfortunately, a relative error of ~ 50 % should be considered for all mixing ratios for line D, because of a poor SNR. The above mentioned bandwidth effects reduced the peak absorption even further to just above noise level and thus uncertainties in the baseline polynomial fits gain more influence.

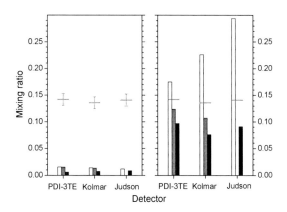

Figure 4.23: CH_4 mixing ratios deduced from lines C (left panel) and D (right panel) using the integrated absorption coefficient at $t_{pulse} = 1.5$ ns (blank), the time-integrated detector signal (grey) or the result from fit of *TDL Wintel* (black). Mixing ratios of the simultaneous TDLAS measurements are indicated by lines. (For detector parameters see table 4.3; $U_{QCL} = 11.7$ V).

The present comparison of spectra and concentration values demonstrates that the frequency chirp of pulsed QCLs is only a part of the explanation for observed obstacles such as distorted line shapes or underestimated molecular mixing ratios. It is also essential to take bandwidth effects of the detection system (i.e. detector, preamplifier, digitiser) or the detection approach (i.e. time-integrated or time resolved signals) into account. For the resulting spectrum and hence the calculated concentrations the maximum effective bandwidth of the system is a critical parameter. On the one hand it may be limited by integrating over the chirped laser pulse as achieved with gated integrators [72,104] or in a post-measurement data analysis as presented here. On the other hand the effective bandwidth is often limited by the detector type and element itself (e.g. LN cooled with standard size of 1 mm^2) or the subsequent electronics. If the effective bandwidth falls below a cut-off value which is determined by the chirp rate of the laser, rapid passage structures at low pressure conditions and the chirped pulse are blurred in the recorded spectra. The apparent spectral broadening is often considered as effective laser line width of pulsed QCLs.

Carefully chosen operation parameters for the laser might give correct results with the *inter* pulse method at elevated pressures (section 4.2.3) where pressure broadening approaches the instrumental broadening or effective laser line width. In contrast, at low pressure conditions the laser chirp and the effective bandwidth of the system, typically limited by the digitiser, cause a quasi-multimode behaviour leading to reduced and distorted absorption features. Thus, fit procedures using symmetric line profiles or the integrated absorption coefficient yield wrong mixing ratios. These effects are more pronounced at higher laser intensities, i.e. clearly above the threshold, and for strong absorption features. In this case non-linear absorption phenomena are present, e.g., rapid passage and power saturation. Due to bandwidth limitation these effects are hidden in the acquired spectra and cannot be identified accurately.

4.5   Discussion and conclusions

Recently, IRLAS using a new type of semiconductor lasers, namely QCLs, has been successfully applied especially in the field of gas phase spectroscopy. In many cases the measurements were performed at elevated pressures (> 50 mbar) which is a compromise between spectral selectivity and sensitivity. Particularly, the latter figure of merit of QCL based spectrometers is degraded by the specific behaviour of frequency swept QCLs accompanied by non-linear absorption effects. It is known and was confirmed here that these challenges play a minor role at elevated pressures, but their origin and the consequences at lower pressures (< 10 mbar) is not yet fully understood. For this reason, the same spectrometer which gave reasonable results at 100 mbar was also used in the low pressure range. *In-situ* measurements in low pressure plasmas as an essential part of this thesis require a better understanding of the observed phenomena and approaches to obviate them. The study mainly focussed on tuning methods which enable a time resolution below 1 s to be accomplished which is not of main interest for trace gas measurements but crucial for plasma processes on the micro- or millisecond time scale. Hence the *inter* pulse mode, directly deduced from TDLAS where a current ramp is impressed on the laser, and the *intra* pulse mode based on the high current pulse induced temperature and frequency chirp of the laser were scrutinised.

The present analysis showed that the short pulse mode exhibits all characteristics of the *intra* pulse mode, i.e. the frequency chirp of the QCL is always present, but the principle of this tuning method introduces further complications if applied to QCLs instead of TDLs. Time resolved studies of the short pulses and a quantitative comparison with results from a co-aligned lead salt laser system established asymmetrical instrumental broadening and up to one order of magnitude underestimated concentrations at low pressure conditions. A detailed analysis revealed a complex combination of three interdependent effects:

 i) the frequency chirp of the QCL,
 ii) bandwidth limitations of the entire spectrometer,
 iii) non-linear absorption behaviour.

The analogue bandwidth of the entire system (ii) is chirp rate dependent and an important criterion, but typically neglected in the design of spectrometers .

It was found for more than 30 different QCLs that the average chirp rate for longer pulses never falls below a lower limit of 0.003 cm^{-1}/ns (90 MHz/ns). However, for short pulses, as applied in the *inter* pulse mode, the chirp rate (dν/dt) is often higher in the order of at least 0.005 cm^{-1}/ns (150 MHz/ns). Even for extremely short pulses the total laser chirp can no longer be considered as small or negligible compared to typical Doppler broadened absorption lines at low pressures. Very short optical pulses of 1 - 2 ns are only accomplished at the laser threshold exhibiting a limited output power and enhanced pulse-to-pulse fluctuations which reduce the sensitivity in any case. The analysis of all laser pulse widths during a frequency scan showed an increase at higher laser frequencies, but the observed optical pulses were still clearly shorter than the nominal electrical pulse width. Since the QCL is operated close to its threshold the current exceeds this value only for a time window shorter than the nominal width and so is the optical pulse width. Due to the negative temperature tuning coefficient higher laser frequencies are achieved at lower temperatures. The sub-threshold current is thus ramped such that the laser temperature decreases which simultaneously reduces the threshold. This causes increased QCL pulses at the end of the sweep. A better SNR with this method is only accomplished at higher operation voltages (or currents) which also increases the laser pulse width, the chirp rate and thus the chirp. Furthermore, the effective laser line width rises. As long as a single pulse, i.e. a single measurement channel in terms of the *inter* pulse principle, is considered the laser line $\Delta\nu$ width is approximated by Duxbury's formula ($\Delta\nu \cdot c \sim (C \cdot d f/d t)^{1/2}$) where the laser line width is directly connected to the chirp rate. Even for the lower limit of the chirp rates for short pulses the laser line width of a single pulse (FWHM) exceeds 0.008 cm^{-1} (250 MHz) and therefore the typical Doppler broadening.

Due to the measurement principle of the *inter* pulse mode the laser chirp of a single pulse is spread over several channels which was shown with a highly time resolved acquisition of individual pulses and a post-measurement analysis and deconvolution. As soon as the bandwidth of the entire system is limited, either artificially by time-integration of the detector signal containing the chirped pulse or naturally by the detection and digitising system, the laser chirp appears as an asymmetrical spectral broadening. Absorption lines, particularly at low pressures, may exhibit a shadow-like feature on their low or high frequency tail. It can be considered as a quasi-multimode behaviour due to the chirp since the peak absorption is also reduced. It is often described by an effective laser line width which is typically much higher (> 0.02 cm^{-1}, 600 MHz, FWHM) than estimated above, but it is in fact a consequence of the measurement principle and the limited bandwidth of the spectrometer. These artefacts become more pronounced at the end of the laser sweep because the laser pulse width and the chirp increases.

Although the oscillatory structure of the rapid passage effect is present in the deconvoluted spectra, the bandwidth limited *inter* pulse graphs show no such features. Thus, quantitative results at low pressure not only suffer from the quasi-multimode behaviour including reduced absorption, but also from all non-linear absorption effects observed with the *intra* pulse mode. Especially the rapid passage effect is more pronounced since the chirp rate is higher for short pulses. The smallest discrepancies between expected and measured concentrations are observed for weak absorption features which leads to the same consequences as in the *intra* pulse mode.

It is interesting to discuss common approaches and suggestions for using the *inter* pulse mode in respect to the presently obtained results. Typically the short pulse method is used with total pressures above 50 mbar where the rapid passage effect is damped, power saturation is also reduced and occurs only for very strong absorption features ($S \geq 10^{-20}$ cm/molecule for main isotopes) and the pressure broadening approaches the instrumental broadening. In this case correct mixing ratios may be deduced, because laser chirp and bandwidth effects play a minor role. However, the experimental conditions for *in-situ* measurements in low pressure plasmas are clearly different. Additionally, laser operation close to the threshold and short laser pulses generated with pulse electronics of short rise time are suggested [22] and should reduce the QCL chirp. Reasonable results employing a fit procedure may be achieved in the case of weak absorption (≤ 1 %) and with an instrumental broadening as a calibration factor set to the effective laser line width. This approach only minimises the systematic errors, but cannot overcome the general shortcomings of the method. Bandwidth effects are taken into account by an arbitrary calibration of the instrumental broadening. It also reduces the SNR due to a simultaneously reduced laser intensity close to the threshold and thus hampers the application of multi-pass cells or even optical cavities where higher optical power is required to balance reflection losses.

Only recently, bandwidth effects have been mentioned in an independent investigation of potential drawbacks of using pulsed QCLs [54]. Grouiez et al. used an experimental arrangement where the bandwidth of all individual components was high enough to resolve the internal rapid passage structure of *inter* pulse spectra. They followed a similar approach as presented here to separate the frequency-down chirp during the laser pulse, which represents already a portion of a spectrum, from the laser sweep. In both studies a constant t_{pulse} was monitored during the frequency scan which was achieved by modulating the heat sink temperature in their case. Unfortunately, it was concluded that the non-linear effects could be avoided by placing the absorption line at the end of a longer (so called intermediate) pulse and tried to confirm this with measurements at 60 mbar. The rapid passage is definitely reduced due to a decrease of the chirp rate, but it was finally damped as a consequence of the relatively high pressure. Since the chirp rate of pulsed QCLs is typically sufficiently high to generate rapid passage effects for molecular transitions in the MIR (normalised sweep rate $A \gg 300 \gg 1$), it is essential to consider the (linear) rapid passage regime to be always present at low pressures.

It is more interesting that the authors propose bandwidth effects as a reason for the distorted line shapes in the *inter* pulse mode. The present results indicate that any limitation to the analogue bandwidth of the system caused by the detector element, detector type, preamplifier, digitiser or the measuring principle (e.g. integrating over the optical pulse) appears as a strong instrumental broadening in the spectra. A relation to estimate the minimum bandwidth BW_{min} required for reproducing the fast transients in the signals was determined ($BW_{min} \sim (0.49 \cdot C^{-1} \cdot \mathrm{d}f/\mathrm{d}t)^{1/2}$) and found to be chirp rate dependent. Generally, the analogue bandwidth should not fall below 250 MHz. Provided that the entire system is designed for such high analogue bandwidths the pulse-to-pulse jitter between subsequent pulses (~ 2 ns) was found to remain a challenge which introduces additional noise. In many cases an additional reference channel is necessary for triggering purposes [30,31] and to reduce pulse-to-pulse intensity fluctuations with a pulse normalisation [22,23].

It is therefore obvious that a straightforward replacement of the light source and in some cases of the detector in conventional TDLAS spectrometers is not sufficient to accomplish a calibration free QCLAS spectrometer. The digitiser is normally adapted to the ≤ 1 MHz LN cooled detectors of lead salt laser based systems and would now limit the bandwidth and blur non-linear absorption effects. The resulting effective laser line width is thus a complex and inappropriate figure of merit encompassing several other individual laser and system parameters, among them, chirp rate, pulse length, current ramp parameters etc. Consequently, the reported values (FWHM) range from 0.008 to 0.093 cm^{-1} (0.24 ... 2.79 GHz, table 4.4).

The term laser line width should be carefully used and may be classified as follows. The intrinsic spectral width of cw QCLs is about (FWHM) 4×10^{-7} cm^{-1} (12 kHz) [102], but it is typically not observed due to current instabilities of the controllers which limit the line width to 3×10^{-5} ... 0.0013 cm^{-1} (1 ... 40 MHz) in practice [e.g., 70,94,99]. The line width of pulsed QCLs is determined by the uncertainty relation and can be either Fourier or chirp rate limited. The first case is observed for extremely short pulses < 5 ns and yields laser line widths (FWHM) > 0.005 cm^{-1} (150 MHz). Longer pulses are chirp rate limited and spectral widths (FWHM) may be estimated for a best aperture time to ~ 0.010 cm^{-1} (300 MHz, $C = 0.886$) and ~ 0.007 cm^{-1} (210 MHz, $C = 0.441$) respectively. The full chirp of an at least 5 ns long pulse would be ≥ 0.02 cm^{-1} (600 MHz). In the case of bandwidth limited systems the effective laser line width is found between the chirp rate limit and the total chirp of the laser and depends on the operation conditions, especially on the pulse width.

Several spectrometers using the *inter* pulse mode are reported in the literature and are collected with their main parameters in table 4.4. The bandwidth limit and the potential bottleneck are marked. In some cases the detector-preamplifier combination sets the limit [50,58,112]. This might be due to the fact that LN cooled detectors with their inherently higher sensitivity were preferred at the expense of a slower response time. In several studies the digitiser (e.g., DAC) adapted to former TDLAS setups with low bandwidth detectors finally limited the system performance and integrated over the full chirp [22,23,53,57]. The current results and the experiments of Grouiez et al. [54] clearly show that an appropriate analogue bandwidth and measuring principle may yield the fundamental laser line width (either chirp rate or Fourier limited). The comparison in table 4.4 also includes an FTIR study where the emission of a pulsed QCL was directly recorded [67]. Although the absolute values may be limited by the spectrometer resolution, a difference between chirp rate and Fourier limited pulses was visible.

Since particularly the rapid passage effect cannot be obviated under low pressure conditions a calibration of the *inter* pulse mode is always required and does not provide any further advantage compared to the *intra* pulse method. In respect to the available tuning range and the achievable time resolution the latter mode of operation is even superior. Hence the short pulse mode is not considered for linear absorption spectroscopic studies of low pressure molecular plasmas in this thesis.

Effective Laser Line Width (FWHM) [cm⁻¹]	[MHz]	Pulse Width [ns]	Band-width Limit [MHz]	Fourier[a]	Chirp Rate	Det.[b]	PA[c]	DAC[d]	Total Pressure [mbar]	Ref.
0.024	720	11	?			LN		?	53	[1]
0.010	290	5	~ 20	(X)[e]		LN	X	X	~ 2	[49]
0.099	3000	50	20			LN	X		64	[50]
0.020	600	5	~ 20			LN		X[f]	126	[72]
0.020	600	13	1			TE		X	48	[22]
0.040	1200	20	?			TE	?	?	1013	[105]
0.012	360	15	[g]1			LN	X	X	~ 20	[53]
0.020	600	18	[g]1			LN	X	X	~ 20	[53]
0.024	720	10...15	1			TE		X	48	[23]
0.020	600	10...15	1			TE		X	60	[23]
0.008	230	10...15	1			TE		X	50	[23]
0.008	250	15	< 50			LN		X[f]	53	[104]
~ 0.050	1500	25	1			LN		X	60	[103]
0.040	1200	25	50			TE	X		67	[h][58]
0.024	720	10	5			LN	X		133	[i][112]
0.024	720	10	1			TE		X	64	[57]
0.008	240	12	1			TE		X	0.06...3	[25]
0.020	600	5...10	?			TE		X	266	[59]
-	-	7	350	X				X	0.5	[54]
[j]0.013		7	0...90					X	0.5	[54]
[j]0.013		7	120			X LN			0.6	[54]
-	*-*	*0...10*	*600*	*X*		*TE*	*X*		*0.5*	*This*
0.012	*360*	*0...10*	*5*			*X LN*			*0.5*	*...*
0.012	*360*	*0...10*	*1*			*LN/TE*		*X*	*0.5*	*work*
0.048	1440	3	?	(X)[k]		FTIR	X	X	-	[67]
0.048	1440	10	?		(X)[k]	FTIR	X	X	-	[67]
> 0.050	1500	> 10	?			FTIR	X	X	-	[67]

Table 4.4: Intercomparison of effective laser line widths of pulsed QCLAS spectrometers employing the short pulse mode. The potential origin of the instrumental broadening is marked with X. A FT-IR study of the direct emission of a pulsed QCL is added. (All pressures were converted into mbar.)

[a] Fourier limited according to the uncertainty relation
[b] detector
[c] preamplifier
[d] digitiser or DAC equipment
[e] probably Fourier limited, but detection bandwidth was too low
[f] gated integrator used
[g] bandwidth of all components matched 1 MHz
[h] temperature tuning used instead of a current ramp
[i] observed in combination with an optical cavity
[j] estimated from the reported spectra
[k] additionally broadened by the FTIR resolution, but clearly visible

The application of the *intra* pulse mode facilitates a time resolution in the order of hundreds of microseconds if the intrinsic repetition frequency of the QCL is used or even down to the pulse length of ~ 100 ns if special trigger schemes are employed. The recorded spectra typically suffer from 4 non-linear absorption phenomena

i)   rapid passage effects due to the rapidly frequency chirped laser,

ii) power saturation caused by the high laser intensities,

iii) self-focussing of the laser beam due to the high output power [55],

iv) molecular alignment as a consequence of the linearly polarised QCL emission [34].

At elevated pressures the generalised relaxation rate $\gamma = (T_1 T_2)^{1/2}$ (eqs. (2 - 33), (2 - 34)), being several hundred MHz as estimated from the homogeneous broadening $\Delta f_{col} \approx \gamma$ (eqs. (2 - 9), (2 - 11) and (2 - 26)), efficiently suppresses these effects and undisturbed spectra are observed. However at low pressures (< 10 mbar) the relaxation rate or homogeneous broadening drops to tens of MHz. The relaxation time $(1/\gamma)$ is then in the order of 100 ns and hence much longer than the typical interaction time of the chirped laser ($\geq$ 100 MHz/ns) with the homogeneously broadened transition (Δf_{col} ~ 10 MHz)[7] which would be several hundred picoseconds. Consequently, rapid passage effects are easily accomplished [55]. Typical chirp rates of more than 30 MHz/ns (0.001 cm^{-1}/ns) were found for many lasers in the current study. In contrast, power saturation might be present only for strong absorption features as illustrated here for CO and CH$_4$. The analysis of the experiments demonstrated that calculations employing the nominal laser intensity and beam waist typically do not yield the corresponding saturation parameter (presently especially valid for CO). Self-focussing of the laser beam may occur to accomplish the required power densities for the saturation [55].

To summarise, low pressure spectra acquired with pulsed QCLs are affected by non-linear absorption phenomena, i.e. Beer-Lambert's law is violated. All details in this respect are still not entirely understood and a theoretical treatment is beyond the scope of this thesis work. In particular pulsed QCLs should be therefore not considered as an equivalent substitute to lead salt lasers as MIR light source. Fast passage, power saturation and molecular alignment generally require a complicated and time consuming numerical calculation. In practice alternative data analysis approaches are thus necessary to obviate determining underestimated molecular concentrations.

The systematic error can be reduced by means of several measures. Weak absorption features and multiple pass optics are desirable to reduce potential power saturation and sustain a reasonable sensitivity of the spectrometer. Additionally, the absorption feature of interest should be shifted to the end of the laser pulse where the chirp rate and thus the rapid passage effect is often reduced. Provided no power saturation occurs a new data analysis approach can successfully be applied: partial integration over the undisturbed half of the absorption line yielded correct mixing ratios. In other words, the systematic error induced by this approximation is kept below the measurement uncertainties. If stronger absorption features are considered, a straightforward calibration is suggested yielding a correction factor for defined experimental conditions [e.g., 26 and chapter 5,27,30,31].

---

[7]   At low pressure conditions Doppler-broadening clearly exceeds the homogeneous pressure broadening. However, the relaxation is still determined by the collisions whereas the inhomogeneous Doppler broadening must be treated separately [55].

B Appendix

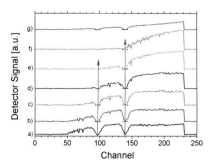

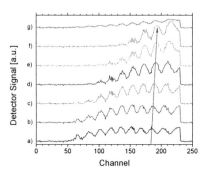

Figure B.01: CH_4 absorption spectra (10 mbar, 0.15 m) determined from the detector signal at $t_{pulse} = $ const. during a chirped QCL pulse: a) 1.0 ns, b) 1.5 ns, c) 2.0 ns, d) 2.5 ns, e) 3.0 ns, f) 3.5 ns, g) averaged signal over the entire pulse. The zero level from the gate-off period is indicated by dotted lines. Arrows represent the shift of spectral features with t_{pulse}. ($U_{QCL} = 11.5$ V = threshold).

Figure B.02: Etalon spectra (FSR = 0.048 cm^{-1}) determined from the detector signal at $t_{pulse} = $ const. during a chirped QCL pulse: a) 1.0 ns, b) 1.5 ns, c) 2.0 ns, d) 2.5 ns, e) 3.0 ns, f) 3.5 ns, g) averaged signal over the entire pulse. The shift of the extrema with t_{pulse} is marked with an arrow. ($U_{QCL} = 11.5$ V = threshold).

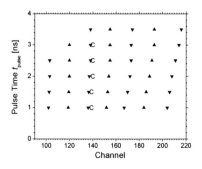

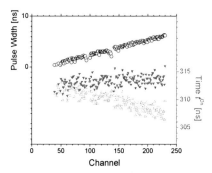

Figure B.03: Shift of the position of maxima (▲) and minima (▼) of the deconvoluted etalon spectra (figure B.02) and of the strong CH_4 line C (feature B was not analysed) in figure B.01. ($U_{QCL} = 11.5$ V = threshold).

Figure B.04: Start (△) and end (▼) point of the QCL pulse on the time scale t^{Ch} (right hand scale, grey) and the corresponding difference yielding the pulse width (○, left hand scale, black). ($U_{QCL} = 11.5$ V = threshold).

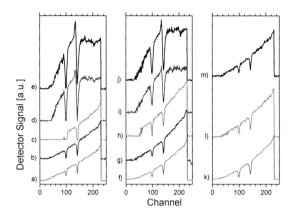

Figure B.05: CH$_4$ spectra (0.5 mbar, mixing ratio = 0.14, 24 m) measured by using a detector/preamplifier with an effective bandwidth of 600 MHz (left), 50 MHz (middle) and 5 MHz (right). The oscilloscope bandwidth was 20 MHz (c, h, l), 200 MHz (d, i) and 1 GHz for all other traces. Traces (c, d, e) and (h, i, j) represent the detector signal at t_{pulse} = 1.5 ns whereas (b, g) show the time-integrated signal; (l, m) were taken at t_{pulse} = 22 ns ($\sim$ maximum) due to the slow rise time of the detector. Additionally, the detector signals were measured with the *TDL Wintel* system (a, f, k). (U_{QCL} = 11.5 V).

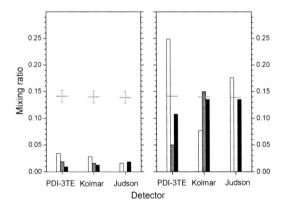

Figure B.06: CH$_4$ mixing ratios deduced from lines C (left panel) and D (right panel) using the integrated absorption coefficient at t_{pulse} = 1.5 ns (blank), the time-integrated detector signal (grey) or the result from fit of *TDL Wintel* (black). Mixing ratios of the simultaneous TDLAS measurements are indicated by lines. (For detector parameters see table 4.3; U_{QCL} = 11.5 V = threshold).

Bibliography

[1]   K. Namjou, S. Cai, E.A. Whittaker, J. Faist, C. Gmachl, F. Capasso, D.L. Sivco, A.Y. Cho, *Opt. Lett.* **23**, 219 (1998).

[2]   E. Normand, M. McCulloch, G. Duxbury, N. Langford, *Opt. Lett.* **28**, 16 (2003).

[3]   T. Beyer, M. Braun, A. Lambrecht, *J. Appl. Phys.* **93**, 3158 (2003).

[4]   C. Gmachl, F. Capasso, R. Köhler, A. Tredicucci, A.L. Hutchinson, D.L. Sivco, J.N. Baillargeon, A.Y. Cho, *IEEE Circ. Dev.* **16**, 10 (2000).

[5]   A.A. Kosterev, F.K. Tittel, *IEEE J. Quant. Electron.* **38**, 582 (2002).

[6]   L. Hvozdara, S. Gianordoli, G. Strasser, W. Schrenk, K. Unterrainer, E. Gornik, C.S.S.S. Murthy, M. Kraft, V. Pustogow, B. Mizaikoff, A. Inberg, N. Croitoru, *Appl. Opt.* **39**, 6929 (2000).

[7]   G. Gagliardi, F. Tamassia, P. De Natale, C. Gmachl, F. Capasso, D.L. Sivco, J.N. Baillargeon, A.L. Hutchinson, A.Y. Cho, *Eur. Phys. J. D* **19**, 327 (2002).

[8]   G. Gagliardi, S. Borri, F. Tamassia, F. Capasso, C. Gmachl, D.L. Sivco, J.N. Baillargeon, A.L. Hutchinson, A.Y. Cho, *Isot. Environ. Health Stud.* **41**, 313 (2005).

[9]   D. Weidmann, C.B. Roller, C. Oppenheimer, A. Fried, F.K. Tittel, *Isot. Environ. Health Stud.* **41**, 293 (2005).

[10]   J.B. McManus, D.D. Nelson, J.H. Shorter, R. Jimenez, S. Herndon, S. Saleska, M. Zahniser, *J. Mod. Opt.* **52**, 2309 (2005).

[11]   K.G. Hay, S. Wright, G. Duxbury, N. Langford, *Appl. Phys. B* **90**, 329 (2008).

[12]   M. Taslakov, V. Simeonov, H. van den Bergh, *Spectrochim. Acta Part A* **63**, 1002 (2006).

[13]   J.D. Whitehead, M. Twigg, D. Famulari, E. Nemitz, M.A. Sutton, M.W. Gallagher, D. Fowler, *Environ. Sci. Technol.* **42**, 2041 (2008).

[14]   C.L. Schiller, H. Bozem, C. Gurk, U. Parchatka, R. Königstedt, G.W. Harris, J. Lelieveld, H. Fischer, *Appl. Phys. B* **92**, 419 (2008).

[15]   G. Wysocki, Y. Bakhirkin, S. So, F.K. Tittel, C.J. Hill, R.Q. Yang, M.P. Fraser, *Appl. Opt.* **46**, 8202 (2007).

[16]   C. Bauer, A.K. Sharma, U. Willer, J. Burgmeier, B. Braunschweig, W. Schade, S. Blaser, L. Hvozdara, A. Müller, G. Holl, *Appl. Phys. B* **92**, 327 (2008).

[17]   C. Bauer, U. Willer, R. Lewicki, A. Pohlkötter, A. Kosterev, D. Kosynkin, F.K. Tittel, W. Schade, *J. Phys.: Conf. Ser.* **157**, 012002 (2009).

[18]   J. Hildenbrand, J. Herbst, J. Wöllenstein, A. Lambrecht, *Proc. SPIE* **7222**, 72220B (2009).

[19]   A. Edelmann, C. Ruzicka, J. Frank, B. Lendl, W. Schrenk, E. Gornik, G. Strasser, *J. Chromat. A* **934**, 123 (2001).

[20]   S.M. Cristescu, S.T. Persijn, S. te Lintel Hekkert, F.J.M. Harren, *Appl. Phys. B* **92**, 343 (2008).

[21]   M.R. McCurdy, Y. Bakhirkin, G. Wysocki, R. Lewicki, F.K. Tittel, *J. Breath Res.* **1**, 014001 (2007).

[22]   D.D. Nelson, J.H. Shorter, J.B. McManus, M.S. Zahniser, *Appl. Phys. B* **75**, 343 (2002).

[23]   D.D. Nelson, J.B. McManus, S. Urbanski, S. Herndon, M.S. Zahniser, *Spectrochim. Acta A* **60**, 3325 (2004).

[24]   A. Cheesman, J.A. Smith, M.N.R. Ashfold, N. Langford, S. Wright, G. Duxbury, *J. Phys. Chem. A* **110**, 2821 (2006).
[25]   G.D. Stancu, N. Lang, J. Röpcke, M. Reinicke, A. Steinbach, S. Wege, *Chem. Vap. Deposition* **13**, 351 (2007).
[26]   S. Welzel, L. Gatilova, J. Röpcke, A. Rousseau, *Plasma Sources Sci. Technol.* **16**, 822 (2007).
[27]   J.H. van Helden, S.J. Horrocks, G.A.D. Ritchie, *Appl. Phys. Lett.* **92**, 081506 (2008).
[28]   F. Hempel, V. Artyushenko, F. Weichbrodt, J. Röpcke, *J. Phys.: Conf. Ser.* **157**, 012003 (2009).
[29]   N. Lang, J. Röpcke, H. Zimmermann, A. Steinbach, S. Wege, *J. Phys.: Conf. Ser.* **157**, 012007 (2009).
[30]   S. Welzel, S. Stepanov, J. Meichsner, J. Röpcke, *J. Phys.: Conf. Ser.* **157**, 012010 (2009).
[31]   S. Stepanov, S. Welzel, J. Röpcke, J. Meichsner, *J. Phys.: Conf. Ser.* **157**, 012008 (2009).
[32]   J.T. Remillard, D. Uy, W.H. Weber, F. Capasso, C. Gmachl, A.L. Hutchinson, D.L. Sivco, J.N. Baillargeon, A.Y. Cho, *Opt. Exp.* **7**, 243 (2000).
[33]   A. Castrillo, G. Casa, L. Gianfrani, Opt. Lett. **32**, 3047 (2007).
[34]   G. Duxbury, N. Langford, M. McCulloch, S. Wright, *Mol. Phys.* **105**, 741 (2007).
[35]   J.H. van Helden, R. Peverall, G.A.D. Ritchie, R.J. Walker, *Appl. Phys. Lett.* **94**, 051116 (2009).
[36]   B.A. Paldus, C.C. Harb, T.G. Spence, R.N. Zare, C. Gmachl, F. Capasso, D.L. Sivco, J.N. Baillargeon, A.L. Hutchinson, A.Y. Cho, *Opt. Lett.* **25**, 666 (2000).
[37]   M.S. Taubman, T.L. Myers, B.D. Cannon, R.M. Williams, *Spectrochim. Acta Part A* **60**, 3457 (2004).
[38]   S. Bartalini, S. Borri, I. Galli, D. Mazotti, P.C. Pastor, G. Giusfredi, P. De Natale, *Proc. SPIE* **7222**, 72220C (2009).
[39]   S. Borri, S. Bartalini, I. Galli, P. Cancio, G. Giusfredi, D. Mazzotti, A. Castrillo, L. Gianfrani, P. De Natale, *Opt. Exp.* **16**, 11637 (2008).
[40]   G. Wysocki, R. Lewicki, R.F. Curl, F.K. Tittel, L. Diehl, F. Capasso, M. Troccoli, G. Hofler, D. Bour, S. Corzine, R. Maulini, M. Giovannini, J. Faist, *Appl. Phys. B* **92**, 305 (2008).
[41]   N. Mukherjee, C.K.N. Patel, *Chem. Phys. Lett.* **462**, 10 (2008).
[42]   Q. Wen, K.H. Michaelian, *Opt. Lett.* **16**, 1875 (2008).
[43]   M.C. Phillips, N. Ho, *Opt. Exp.* **16**, 1836 (2008).
[44]   F. Capasso, R. Paiella, R. Martini, R. Colombelli, C. Gmachl, T.L. Myers, M.S. Taubman, R.M. Williams, C.G. Bethea, K. Unterrainer, H.Y. Hwang, D.L. Sivco, A.Y. Cho, A.M. Sergent, H.C. Liu, E. A. Whittaker, *IEEE J. Quantum Electron.* **38**, 511 (2002).
[45]   J. Faist, D. Hofstetter, M. Beck, T. Aellen, M. Rochat, S. Blaser, *IEEE J. Quantum Electron.* **38**, 533 (2002).
[46]   R. Martini, C. Bethea, F. Capasso, C. Gmachl, R. Paiella, E.A. Whittaker, H.Y. Hwang, D.L. Sivco, J.N. Baillargeon, A.Y. Cho, *IEEE El. Lett.* **38**, 181 (2002).
[47]   G. Cheng, C.G. Bethea, R. Martini, presented at the 9th Int. Conf. on Mid-Infrared Optoelectronics: Materials and Devices (MIOMD), Freiburg, Germany, 7 - 11 September 2008.

[48]   N. Yu, E. Cubukcu, L. Diehl, M.A. Belkin, K.B. Crozier, F. Capasso, D. Bour, S.
       Corzine, G. Höfler, *Appl. Phys. Lett.* **91**, 173113 (2007).
[49]   A.A. Kosterev, F.K. Tittel, C. Gmachl, F. Capasso, D.L. Sivco, J.N. Baillargeon,
       A.L. Hutchinson, A.Y. Cho, *Appl. Opt.* **39**, 6866 (2000).
[50]   D.M. Sonnenfroh, W.T. Rawlins, M.G. Allen, C. Gmachl, F. Capasso, A.L.
       Hutchinson, D.L. Sivco, J.N. Baillargeon, A.Y. Cho, *Appl. Opt.* **40**, 812 (2001).
[51]   D.D. Nelson, M.S. Zahniser, J.B. McManus, C.E. Kolb, J.L. Jimenez, *Appl. Phys. B*
       67, 433 (1998).
[52]   M.S. Zahniser, D.D. Nelson, J.B. McManus, P.L. Kebabian, *Phil. Trans. R. Soc.
       Lond. A* **351**, 371 (1995).
[53]   Q. Shi, D.D. Nelson, J.B. McManus, M.S. Zahniser, M.E. Parrish, R.E. Baren, K.H.
       Shafer, C.N. Harward, *Anal. Chem.* **75**, 5180 (2003).
[54]   B. Grouiez, B. Parvitte, L. Joly, V Zeninari, *Opt. Lett.* **34**, 181 (2009).
[55]   G. Duxbury, N. Langford, M.T. McCulloch, S. Wright, *Chem. Soc. Rev.* **34**, 921
       (2005).
[56]   M.T. McCulloch, G. Duxbury, N. Langford, *Mol. Phys.* **104**, 2767 (2006).
[57]   J.B. McManus, D.D. Nelson, S.C. Herndon, J.H. Shorter, M.S. Zahniser, S. Blaser,
       L. Hvozdara, A. Muller, M. Giovannini, J. Faist, *Appl. Phys. B* **85**, 235 (2006).
[58]   D. Weidmann, G. Wysocki, C. Oppenheimer, F.K. Tittel, *Appl. Phys. B* **80**, 255
       (2005).
[59]   J. Manne, W. Jäger, J. Tulip, *Appl Phys B* **94**, 337 (2009).
[60]   M.S. Zahniser, D.D. Nelson, J.B. McManus, S.C. Herndon, E.C. Wood, J.H. Shorter,
       B.H. Lee, G.W. Santoni, R. Jimenez, B.C. Daube, S. Park, E.A. Kort, S.C. Wofsy,
       Proc. SPIE **7222**, 72220H (2009).
[61]   B.M. Gorshunov, A.P. Shotov, I.I. Zasavitsky, V.G. Koloshnikov, Y.A. Kuritsyn,
       G.V. Vedeneeva, *Opt. Comm.* **28**, 64 (1979)
[62]   J. Röpcke, L. Mechold, M. Käning, J. Anders, F.G. Wienhold, D. Nelson, M.
       Zahniser, *Rev. Sci. Instr.* **71**, 3706 (2000).
[63]   J.B. McManus, D. Nelson, M. Zahniser, L. Mechold, M. Osiac, J. Röpcke, A.
       Rousseau, *Rev. Sci. Instr.* **74**, 2709 (2003).
[64]   F. Capasso, J. Faist, C. Sirtori, A.Y. Cho, *Sol. State Comm.* **102**, 231 (1997).
[65]   A.Y. Cho, D.L. Sivco, H.M. Ng, C. Gmachl, A. Tredicucci, A.L. Hutchinson, S.G.
       Chu, F. Capasso, *J. Cryst. Growth* **227-228**, 1 (2001).
[66]   C. Gmachl, F. Capasso, A. Tredicucci, D.L. Sivco, J.N. Baillargeon, A.L.
       Hutchinson, A.Y. Cho, *Opt. Lett.* **25**, 230 (2000).
[67]   D. Hofstetter, M. Beck, J. Faist, M. Nägele, M.W. Sigrist, *Opt. Lett.* **26**, 887 (2001).
[68]   F. Fuchs, B. Hinkov, C. Wild, Q.K. Yang, W. Bronner, K. Köhler, J. Wagner,
       presented at the 9th Int. Conf. on Mid-Infrared Optoelectronics: Materials and
       Devices (MIOMD), Freiburg, Germany, 7 - 11 September 2008.
[69]   F. Fuchs, *Remote Detection of Explosives Using Infrared Semiconductor Lasers*, in:
       IAF Fraunhofer *Semiconductor Lasers and Light Emitting Diodes*, Annual Report,
       2007.
[70]   S.W. Sharpe, J.F. Kelly, J.S. Hartman, C. Gmachl, F. Capasso, D.L. Sivco, J.N.
       Baillargeon, A.Y. Cho, *Opt. Lett.* **23**, 1396 (1998).
[71]   A.A. Kosterev, R.F. Curl, F.K. Tittel, R. Köhler, C. Gmachl, F. Capasso, D.L. Sivco,
       A.Y. Cho, *Appl. Opt.* **41**, 573 (2002).

[72]   A.A. Kosterev, R.F. Curl, F.K. Tittel, R. Köhler, C. Gmachl, F. Capasso, D.L. Sivco,
        A.Y. Cho, S. Wehe, M.G. Allen, *Appl. Opt.* **41**, 1169 (2002).
[73]   V. Spagnolo, A. Lops, G. Scamarcio, M.S. Vitiello, C. Di Franco, *J. Appl. Phys.* **103**,
        043103 (2008).
[74]   H.K. Lee, K.S. Chung, J.S. Yu, *Appl Phys B* **93**, 779 (2008).
[75]   I. Vurgaftman, J.R. Meyer, *J. Appl. Phys.* **99**, 123108 (2006).
[76]   C. Gmachl, A.M. Sergent, A. Tredicucci, F. Capasso, A.L. Hutchinson, D.L. Sivco,
        J.N. Baillargeon, S.N.G. Chu, A.Y. Cho, *IEEE Phot. Technol. Lett.* **11**, 1396 (1999).
[77]   C. Zhu, Y.G. Zhang, A.Z. Li, Z.B Tian, *J. Appl. Phys.* **100**, 053105 (2006).
[78]   A.A. Kosterev, R.F. Curl, F.K. Tittel, M. Rochat, M. Beck, D. Hofstetter, J. Faist,
        Appl. Phys. B **75**, 351 (2002)
[79]   J.U. White, *J. Opt. Soc. Am.* **32**, 285 (1942).
[80]   L.S. Rothman *et al.*, *J. Quant. Spectrosc. Radiat. Transfer* **96**, 139 (2005).
[81]   C. Mann, Q. Yang, F. Fuchs, W. Bronner, K. Köhler, J. Wagner, *Techn. Messen* **72**,
        356 (2005). (*in German*)
[82]   O. Werhahn, J. Koelliker-Delgado, D. Schiel, *Techn. Messen* **72**, 396 (2005). (*in
        German*)
[83]   O. Werhahn, J. Koelliker-Delgado, D. Schiel, *PTB-Mitteilungen* **115**, 305 (2005).
[84]   G. Duxbury, *Infrared Vibration-Rotation Spectroscopy. From Free Radicals to the
        Infrared Sky*, ISBN 0-471-97419-6, (John Wiley & Sons, Chichester, 2000)
[85]   C. Sirtori, J. Nagle, *C. R. Physique* **4**, 639 (2003).
[86]   Y.G. Zhang, Y.J. He, A.Z. Li, *Chin. Phys. Lett.* **20**, 678 (2003).
[87]   K.Q. Le, S. Kim, *Phys. Stat. Sol.* A **205**, 392 (2008).
[88]   M.S. Vitiello, G. Scamarcio, V. Spagnolo, *Appl. Phys. Lett.* **92**, 101116 (2008).
[89]   *Alpes Lasers* (http://www.alpeslasers.ch/).
[90]   M.M.T. Loy, *Phys. Rev. Lett.* **32**, 814 (1974).
[91]   R.R. Ernst, *Sensitivity Enhancement in Magnetic Resonance* in: J.S. Waugh
        Advances in Magnetic Resonance. Vol. 2, (Academic Press, New York, 1966).
[92]   B. Grouiez, B. Parvitte, L. Joly, D. Courtois, V. Zeninari, *Appl. Phys. B* **90**, 177
        (2008).
[93]   T.L. Myers, R.M. Williams, M.S. Taubman, C. Gmachl, F. Capasso, D.L. Sivco, J.N.
        Baillargeon, A.Y. Cho, *Opt. Lett.* **27**, 170 (2002).
[94]   H. Ganser, B. Frech, A. Jentsch, M. Mürtz, C. Gmachl, F. Capasso, D.L. Sivco, J.N.
        Baillargeon, A.Y. Cho, W. Urban, *Opt. Comm.* **197**, 127 (2001).
[95]   C.H. Henry, *IEEE Phot. Technol. Lett.* **18**, 259 (1982).
[96]   M. Osinski, J. Buus, *IEEE J. Quant. Electron.* **23**, 9 (1987).
[97]   C. Harder, K. Vahala, A. Yariv, *Appl. Phys. Lett.* **42**, 328 (1983).
[98]   N. Kumazaki, Y. Takagi, M. Ishihara, K. Kasahara, A. Sugiyama, N. Akikusa, T.
        Edamura, *Appl. Phys. Lett.* **92**, 121104 (2008).
[99]   D.D. Nelson, J.B. McManus, S.C. Herndon, J.H. Shorter, M.S. Zahniser, S. Blaser,
        L. Hvozdara, A. Muller, M. Giovannini, J. Faist, *Opt. Lett.* **31**, 2012 (2006).
[100]   D. Weidmann, L. Joly, V. Parpillon, D. Courtois, Y. Bonetti, T. Aellen, M. Beck, J.
        Faist, D. Hofstetter, *Opt. Lett.* **28**, 704 (2003).
[101]   F. Bielsa, A. Douillet, T. Valenzuela, J.P. Karr, L. Hilico, *Opt. Lett.* **32**, 1641 (2007).
[102]   R.M. Williams, J.F. Kelly, J.S. Hartman, S.W. Sharpe, M.S. Taubman, J.L. Hall, F.
        Capasso, C. Gmachl, D.L. Sivco, J.N. Baillargeon, A.Y. Cho, *Opt. Lett.* **24**, 1844
        (1999).

[103]   G. Wysocki, M. McCurdy, S. So, D. Weidmann, C. Roller, R.F. Curl, F.K. Tittel, *Appl. Opt.* **43**, 6040 (2004).

[104]   G. Wysocki, M. McCurdy, S. So, D. Weidmann, C. Roller, R.F. Curl, F.K. Tittel, *Appl. Opt.* **43**, 3329 (2004).

[105]   S. Schilt, L. Thevenaz, E. Courtois, P.A. Robert, *Spectrochimica Acta Part A* **58**, 2533 (2002).

[106]   M.T. McCulloch, E.L. Normand, N. Langford, G. Duxbury, D. A. Newnham, *J. Opt. Soc. Am. B* **20**, 1761 (2003).

[107]   J. Kauppinen, J. Partanen, *Fourier Transforms in Spectroscopy*, ISBN 3-527-40289-6, (Wiley-VCH, Berlin, 2001).

[108]   T. Butz, *Fourier Transformation for Pedestrians*, ISBN 978-3-540-23165-3, (Springer, Berlin, 2006).

[109]   R. Brüggemann, M. Petri, H. Fischer, D. Mauer, D. Reinert, W. Urban, *Appl. Phys. B* **48**, 105 (1989).

[110]   M. Mürtz, M. Schaefer, M. Schneider, J.S. Wells, W. Urban, U. Schiessl, M. Tacke, *Opt. Comm.* **94**, 551 (1992).

[111]   R.E. Baren, M.E. Parrish, K.H. Shafer, C.N. Harward, Q. Shi, D.D. Nelson, J.B. McManus, M.S. Zahniser, *Spectrochim. Acta Part A* **60**, 3437 (2004).

[112]   M.L. Silva, D.M. Sonnenfroh, D.I. Rosen, M.G. Allen, A.O'Keefe, *Appl. Phys. B* **81**, 705 (2005).

[113]   National Instruments, *User Manual* (http://www.ni.com).

[114]   H. Engels, *Oszilloskop-Meßtechnik von A-Z*, ISBN 3-7723-5912-4, (Franzis, München, 1992). (*in German*).

5   Time resolved study of a pulsed DC discharge using quantum cascade laser absorption spectroscopy

5.1   Introduction

Recent concerns about air quality, including NO_x production, have led to increasing research in the field of pollution abatement from gas exhausts. Apart from conventional techniques, such as catalysis, scrubbers and active carbon, the use of electric discharges is a promising technique for toxic gas removal, especially when these gases are present in low concentrations [1-9]. There is of course a wealth of literature dealing with the NO_x problem and only a few examples directly related to the subject of this chapter are discussed here. One of the key issues when studying plasma processing for gas treatment is to make sure that no undesirable by-product results from the process. Among them, NO_x components are readily produced in air plasmas [9,10].

Over the past few years mid infrared (MIR) absorption techniques are increasingly being used for quantifying studies of phenomena related to NO_x production and removal in plasmas. Using high resolution tuneable diode laser absorption spectroscopy (TDLAS) the production of undesirable NO and NO_2 at the removal process of volatile organic compounds has been studied near atmospheric pressure in a pulsed microwave discharge in air. The influence of changing the pulse duration from 25 to 500 μs and the pulse repetition rate from 10 to 500 Hz is reported. Both NO and NO_2 could be measured simultaneously *ex-situ* in an optical multiple pass cell. In contrast to what was expected, the use of short pulses did not lead to an effective curtailment of the NO_x production. It was found, that the NO_x formation depends only on the average power injected into the plasma and not on the pulse duration and repetition rate [11]. These studies have been extended to the kinetics of low pressure pulsed DC discharges in dry air. Again NO and NO_2 were measured simultaneously by TDLAS downstream of the plasma region. It was shown that the concentration of both species depends only on the average current. Analytical calculations led to fair agreement with the experimental results [12].

The disadvantage of downstream experiments, being separated from the plasma region and being naturally limited in time resolution, have been recently overcome also in DC discharges. TDLAS has been applied *in-situ* in a pulsed low pressure DC discharge of dry air. Under these experimental conditions a time resolution of about 1 ms could be achieved, which was an important step for analysing plasma chemical phenomena in single discharge pulses. It was found that the NO concentration is approximately proportional to the product of the pulse current and the pulse duration. The role of vibrationally excited nitrogen molecules in NO formation was discussed. Numerical computation of a simplified kinetic model for NO formation, taking into account the $N_2(A)$ excited metastable state, showed good agreement [13]. The same NO formation mechanism has been identified by Pintassilgo et al. [14] where $N_2 - O_2$ microwave discharges were studied and modelled for sterilisation purposes. Some other models of $N_2 - O_2$ discharges dealing with NO formation mechanisms in the plasma or the afterglow can be found in the literature. [10,15-22]. In a few cases also surface processes are taken into account which have recently attracted considerable interest. Several investigations in expanding thermal or microwave discharges attempted to discriminate gas phase and surface reactions involved in the NO formation (see also chapter 3) [23,24].

Among other techniques, such as mass spectrometry, *in-situ* TDLAS was employed in double pass or multiple pass cell configurations yielding a time resolution between 0.1 ... 1 s which is sufficient for surface related processes. It was reported in a paper by McManus et al. that with TDLAS systems even higher time resolution can be achieved [25]. Rapid scan software with real time line shape fitting can provide a time resolution to the sub-ms time scale to study chemical kinetic processes of infrared active compounds in plasmas. A higher time resolution, up to the ns scale, could open up a new approach for studying kinetic phenomena in molecular plasmas in real time and *in-situ*. However, the extension of real time measurements on a sub-µs scale to the MIR has been hampered due to the lack of suitable light sources. The feasibility of achieving such a high time resolution, even with tuneable lead salt lasers as MIR source, has been demonstrated by exploiting the internal frequency chirp of a rapidly swept laser [26].

Recent improvements of MIR instrumentation, specifically the advent of quantum cascade lasers (QCLs), offer an attractive new option for infrared absorption spectroscopy for chemical sensing in general and plasma diagnostics in particular. Nowadays QCL absorption spectroscopy (QCLAS) has been used to detect e.g. NO as a trace atmospheric constituent [27] and in exhaled breath [28]. Furthermore it has already been successfully applied to study several plasma processes, e.g. microwave and radio frequency (RF) discharges[1]. Using a QCL the scan through an infrared spectrum is commonly accomplished by changing the laser temperature. Due to the high input power deposited in the device a frequency-down chirp is inherently present in the pulsed mode of operation. Scanning in single, longer pulses (*intra pulse method*) results in an entire spectrum. Since this scan is performed in tens up to a few hundred nanoseconds a time resolution below 100 ns has become possible for quantitative *in-situ* measurements of molecular concentrations in plasmas. Therefore it fits very well to measurements of rapidly changing chemical processes.

Based on this new approach for fast *in-situ* plasma diagnostics an experiment to study the time decay of NO in single discharge pulses of 1 ms and 100 ms duration has been designed. At the centre of interest was the kinetics of the destruction of NO in pulsed DC discharges formed in Ar and N_2 containing mixtures with small admixtures of about 1 % of NO. The comparison of the time evolution of the NO concentration under static and flowing conditions combined with simplified model calculations enabled the analysis of the dynamics of the plasma heating to be achieved. The approach benefits from a significant temperature dependence of the line strength of the measured NO absorption feature. Gabriel et al. considered such influences for the measurement of the CF_2 rotational temperature in pulsed fluorocarbon RF plasmas [29].

The NO time dependence in N_2 - O_2 pulsed discharges has already been measured or calculated using laser induced fluorescence (LIF) [30,31], mass spectrometry [32] or simplified models [32,33]. However, the aims of these previous studies were different from the present case, namely identifying the NO(A) excited state formation [30,31,33] and describing a modulated glow air discharge by a simplified kinetic model [32]. Recently, a kinetic model of a pulsed low pressure DC discharge in N_2 - O_2 (air) has been developed focussing on the NO(X) ground state [34]. Population and depopulation rates can be inferred for different states of the species being included in the model. Main reactions can therefore

---

[1]   See chapters 2 and 4 for a detailed introduction to QCLs, their applications and tuning mechanisms. Appropriate references are also provided there.

readily be identified. The product of the pulse current and duration could be confirmed as the general parameter of pulsed plasmas and the temporal development of N and $N_2(A)$ has been discussed. Calculations on the basis of this model are compared with experimental data of a N_2 (+ NO) discharge.

This chapter is organised as follows: Section 5.2 provides details of the experimental arrangement and in section 5.3 the setup is characterised by means of preliminary experiments. Section 5.4 presents results and a discussion of short and long pulse experiments which are finally summarised in section 5.5.

5.2   Experimental

5.2.1   Discharge setup

The experimental arrangement is shown schematically in figure 5.1 a. The discharge was created in a 50 cm long cylindrical pyrex tube of 2 cm inner diameter (figure 5.1 b). The DC generator, which drives the discharge, was pulsed by a master pulse generator (Philips PM 5715). The time evolution of the discharge current through a resistor and the voltage were measured with high voltage probes (LeCroy PPE20kV, 100 MHz) and a digitising oscilloscope (LeCroy Waverunner 6100, 1 GHz).

A peak current of 35 mA was typically applied to the discharge tube for the duration of 1 ms with a frequency of 1 Hz. In what follows this is referred to as short pulse regime. Additionally, pulse durations of 100 ms (long pulse regime) were achieved with a different DC generator yielding a better stabilised temporal shape of the DC current (19 mA and 35 mA respectively) and thus of the input power. During the pulse the temporal behaviour of the electric field was measured via the voltage difference between two electrodes, (a) and (b) separated by 20 cm. For that purpose the potential difference of each of the electrodes was measured with respect to ground and then subtracted from each other.

Three different types of experiments have been carried out to monitor the decay of NO during   (i)   single (short) plasma pulses under flowing and static gas conditions,
                    i.e. open and closed valves respectively after pre-filling the tube,
            (ii)   a successive number of up to 50 (short) pulses (static conditions only),
            (iii) single (long) pulses, again under flowing and static gas conditions.

The discharge tube was usually pumped down to a base pressure of 0.02 mbar by a rotary pump. For all measurements the total gas pressure was 2 Torr (2.66 mbar), measured with a full range pressure gauge (Pfeiffer PKR251). The gas mixture for case (i) was made of 1 % NO in Ar, with a flow rate of 20 sccm, and of N_2[2], with a flow rate of 2 sccm, giving an initial NO concentration of 0.91 % and a residence time $\tau_{res} = 1.1$ s (eq. 3 - 1)). For the static measurements of experiment (ii) the mixing ratios of Ar/N_2/NO were adjusted to be the same as in (i). The experiments for case (iii) were performed with a standardised gas mixture of 1 % NO in N_2 at a flow rate of 5 sccm resulting in a residence time of almost 5 s. In other

---

[2]   A small fraction of N_2 had to be admixed in order to stabilise the discharge.

words, less than 2 % of the injected molecules (200 ppm or 1.1×10^{13} cm^{-3} of NO) would be renewed during an entire 100 ms plasma pulse. Since this is below the detection limit (see 5.3.2) the gas exchange is negligible during the plasma on-phase.

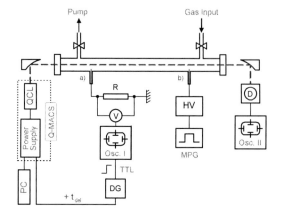

Figure 5.1:

a)
Experimental   setup
(MPG -   master   pulse
generator, DG - delay
generator,   Osc.   -
oscilloscope   I   and   II,
a + b - electrodes used for
potential measurements).

b)
Picture   of   the
experimental setup with
the QCL system (left), the
optical configuration and
the pyrex discharge tube.
Single pass absorption is
indicated   (dotted   line).
The   thermoelectrically
cooled   detector   is   not
shown.

5.2.2   Spectroscopic setup

The QCLAS measurements have been performed using a thermoelectrically (TE) cooled room temperature QCL (Alpes Lasers). The QCL system (Q-MACS, neoplas control) consists of a water cooled laser housing (figure 5.1 b) which prevents thermal drift of the spectral position during the measurements. The heat sink temperature of the laser was kept constant at 23.4 °C (0 °C for (iii)) by means of a two-stage thermoelectric cooler. The QCLAS system used in the present study provides a laser pulse width tuneable between 10 - 255 ns and a repetition frequency between 100 Hz and 1 MHz. The power supply provided a pulsed QCL voltage of 9.6 V of the desired length and was externally triggered to synchronise the measurements with the pulsed discharge. For that a signal had to be used which is directly linked to the breakdown of the discharge and not to the master pulse generator. Thus the delay between the master pulse trigger and the breakdown process as well as the jitter could be omitted. It turned out that the rising edge of the discharge current is steep enough to deliver a stable trigger for all electrical and optical measurements. Once a current pulse had been detected the oscilloscope provided a TTL trigger that was used as an external trigger source for the QCL. The TTL trigger in turn, was not used directly, but via a delay generator (TTI, 10 MHz Pulse Generator TGP110) to probe the plasma pulse as a function of time. The complete trigger scheme is shown in figure 5.2.

In order to sweep the QCL frequency over the NO absorption lines of interest a 90 ns laser pulse (*intra* pulse mode) was used, which was a compromise between having a sufficient frequency scan (i.e. tuning rate) and a reasonable output power of the laser during the pulse. Figure 5.3 shows a typical example. This allows measurements with a time resolution of 90 ns. With the parameters mentioned above the QCL is emitting at 1894 cm^{-1}. In this way the unresolved components of the R 4.5 transition of NO at 1894.151 cm^{-1} and 1894.152 cm^{-1} was observed (figure 5.3). For experiments of case (iii) the QCL heat sink temperature was lowered yielding a higher QCL output power. The laser emitted now at 1897 cm^{-1} corresponding to the unresolved R 5.5 componentes having a line strength comparable to R 4.5 [35].

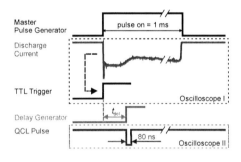

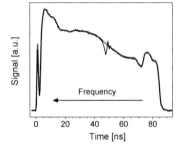

Figure 5.2: Trigger scheme and necessary devices for the time resolved measurements.

Figure 5.3: Spectrum of NO at 1894.15 cm^{-1} (0.9 %, 2.66 mbar, 50 cm) obtained with the *intra* pulse mode of the QCL (thin solid line) superimposed with the raw signal of an empty discharge tube (thick solid line).

The emitted light from the QCL was collimated with a f/0.9 off-axis parabolic (OAP) mirror (38.1 mm focal length) and then directed along the tube axis through the discharge using single pass absorption (experiments (i) + (ii), $L_{eff} = 50$ cm, figure 5.1 b). The transmitted signal was focused with an OAP mirror (31.8 mm focal length) onto a fast TE cooled MCT detector (PDI-3TE series, VIGO System) followed by a fast preamplifier (IPM Freiburg) and a digitising oscilloscope (LeCroy Waverunner LT374L, 500 MHz, figure 5.1 a). The main reason for the mentioned shift in the heat sink temperature was to apply an optical double pass configuration ($L_{eff} = 100$ cm) for all long plasma pulse experiments (iii) to increase the fractional NO absorption. Since in this case the slightly divergent QCL radiation passed the IR transparent KBr tube windows twice often, a higher QCL intensity was desirable.

The data acquisition was fixed to the optical signal by triggering on its falling edge. Hence, the inherent jitter between the driving external trigger for the QCL and the optical pulse, which is typically in the order of 1 ns, can be avoided. For further data analysis the QCL signal without any absorption (baseline) and with a Germanium etalon (25.4 mm, 0.0485 cm^{-1} free spectral range) in the beam path was recorded after each series of measurements.

The single pulse experiments (i) and (iii) have been performed by changing the delay t_{del} between the beginning of the plasma pulse and the trigger for the QCL, in order to measure the NO absorption as a function of time. For the measurements which were carried out under flowing conditions, the signal was always averaged over 5 discharge cycles with the same t_{del}. For static conditions the absorption was recorded during one single laser shot followed by pumping and refilling the tube for the next t_{del} value. In the multi-pulse mode (ii) the delay was fixed to the end of the short plasma pulse ($t_{del} = 900$ μs). The discharge tube was filled once and the NO absorption was then acquired with a single laser shot for each of the successive pulses.

5.3   Characterisation of the arrangement

5.3.1   Electric parameters

The temporal evolution of the electric field E in the discharge under flowing conditions is shown in figure 5.4 for a short plasma pulse. The electric field in the Ar/(N$_2$)/NO mixture decreases gradually from 20 Vcm^{-1} after the breakdown to a steady state of 10 Vcm^{-1} which is reached 400 μs after ignition. Steady state values for our discharge configuration have been calculated from the reduced electric fields E/n ranging from 6 Td to 25 Td in pure argon and pure nitrogen plasmas respectively [36]. The electric field covers then a range between 4 Vcm^{-1} and 16 Vcm^{-1} which agrees well with the experimentally observed values. The breakdown occurs typically within 10 μs as can be seen from the discharge current (figure 5.4) which is of 35 mA during the on-phase of the plasma. The corresponding current density $j = I/A$ with A being the cross sectional area of the discharge tube is then 1.1×10^{-2} Acm^{-2}, assuming a homogeneous current over the tube cross section. The power density jE as a measure for the power injected into the discharge, neglecting other losses, has been deduced from the time resolved measured values of j and E, and is also shown in figure 5.4. It exhibits a peak of 0.3 Wcm^{-3} at the beginning due to the peak in the electric field and levels off at 0.1 Wcm^{-3}. Figure 5.5 depicts similar results for the N$_2$/NO mixture and long plasma pulses under flowing conditions. In this case the current and electric field level off at 38 mA and 26 Vcm^{-1} respectively with a considerable delay after the ignition (~ 20 ms). The reduced electric field would be 40 Td and is thus higher than expected for a pure N$_2$ discharge [36]. Deviations may be caused by overestimating the electric field, since E was calculated from the potential difference across the electrodes without considering sheath losses. Additionally, the small fractions of oxygen in the system may also lead to a slightly increased E/n [36]. However, figure 5.4 also illustrates the reason for choosing this particular power supply: the power density of 0.3 Wcm^{-3} remains almost constant throughout the long plasma pulse.

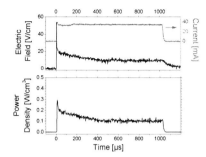 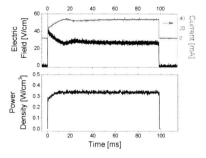

Figure 5.4: Upper: Time resolved measurements of the electric field (black) and of the current through the Ar/(N_2)/NO discharge (grey). **Lower:** Time dependence of the corresponding power density in the plasma. (Values are given for flowing conditions.)

Figure 5.5: Upper: Time resolved measurements of the electric field (black) and of the current through the N_2/NO discharge (grey). **Lower:** Time dependence of the corresponding power density in the plasma. (Values are given for flowing conditions. Note the different time scale to figure 5.4.)

5.3.2   Spectroscopic issues

From the spectroscopic point of view the *intra* pulse mode, which was used in this study to operate the QCL and to facilitate sufficient time resolution, is naturally accompanied by obstacles such as the rapid passage effect - especially at low pressure conditions (figure 5.3, see chapter 4). Power saturation effects may also appear. In order to prevent the influence of these, a calibration procedure has been performed. Due to the relatively low signal-to-noise ratio (SNR), particular for the NO decay in the plasma in single pass configuration, the proposed integration over the undisturbed half of the absorption line (chapter 4, sub-section 4.2.4) is not feasible here.

   Firstly, the recorded raw spectra that are obtained within one single laser pulse have to be converted from the time domain into a wavenumber scale. This is carried out by recording the fringes of a Germanium etalon with a known free spectral range which is shown in figure 5.6. Due to the characteristic response of the QCL to the exciting pulsed current, about 10 ns of the laser signal have to be omitted at each edge of the laser pulse for further analysis. After inverting the time scale as a result of the frequency-down chirp of the laser, the etalon fringes are used to obtain a relative wavenumber calibration. For the QCL used here a slight non-linear correlation between time and the relative frequency was found and fitted with a second order polynomial giving residuals of less than 0.003 cm^{-1}. It turns out that the laser sweep that can be used for measurement purposes is of 60 ns and covers approximately 0.4 cm^{-1}.

   According to equation (2 - 6) the integrated absorption coefficient $K(T)$ has to be evaluated from the spectra to calculate number densities. Therefore, the converted raw spectra are divided by a background spectrum without any absorbing species present yielding the transmission spectrum. After taking the natural logarithm and multiplying it with the inverse absorption length L_{eff} the positive area of the absorption line has been calculated neglecting the artefacts resulting from the rapid passage effect. Finally, the so obtained integrated absorption coefficient $K(T)$ is corrected with a calibration factor which takes account of the simplifications that have been made.

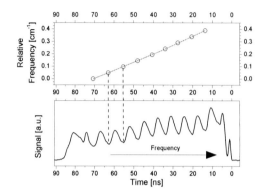

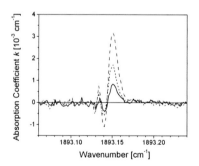

Figure 5.6:
Wavenumber calibration of the 90 ns QCL pulse resulting in a chirp of 0.4 cm^{-1}.

The calibration function is derived from a comparison between the known number densities $n(NO)_{input}$ in several pre-defined gas mixtures and the measured values $n(NO)_{meas}$ on the basis of the integrated absorption coefficients determined as described above. The NO density in the gas mixtures was varied systematically by changing the total gas pressure while keeping the NO mixing ratio constant at 0.91 % (figure 5.7 a) and by changing the mixing ratio while keeping the pressure constant at 2.66 mbar (figure 5.7 b) respectively. The values for $n(NO)_{input}$, covering about one order of magnitude, have been chosen carefully in such a way that all experimentally determined integrated absorption coefficients are considered, i.e. that the calibration is valid for all experimental conditions. The calibration function corresponding to figures 5.7 a and b (measured at 1894 cm^{-1} in single pass configuration) was found to be linear within the entire calibration range, regardless of possible saturation effects (figure 7 in [37]).

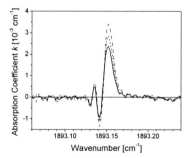

Figure 5.7 a: Sample spectra (absorption coefficient k) for 0.91 % NO admixture at different pressures (solid: 0.665 mbar, dotted: 1.33 mbar, dashed: 2.66 mbar) in single pass configuration.

Figure 5.7 b: Sample spectra (absorption coefficient k) for constant pressure (2.66 mbar) and different NO mixing ratios (solid: 0.67 %, dotted: 0.80 %, dashed: 1.00 %) in single pass configuration.

In this work an alternative calibration (at 1897 cm^{-1} in double pass configuration) is presented (figure 5.8) leading to exactly the same conclusion. The calibration range covers now nearly two orders of magnitude without indications of strong power saturation. The additional calibration results have been selected here, because they enable the detection limit of the (double pass) system to be determined: if averaged spectra (5 times) are considered the NO detection limit is 3×10^{13} cm^{-3} corresponding to 500 ppm NO at 2.66 mbar. Lower number density values simply represent the uncertainties and systematic errors induced by the analysis and calibration as well as the noise level of the detector signal corresponding to about 10^{-2} fractional absorption (see also the discussion in chapter 7). For single shot spectra the noise equivalent fractional absorption increases to 0.03 and so does the NO detection limit. The calibration factor itself which is 1.12 in the present (double pass) configuration exhibits a relative error of typically < 10 %, i.e. the uncertainty induced by the calibration procedure falls below the uncertainty caused by the detector noise.

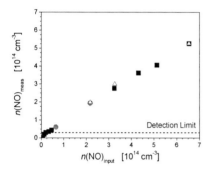

Figure 5.8:
Calibration curve for measured NO absorption in double pass configuration deduced from similar spectra as shown in figures 5.7 a and b (5 times averaged spectra: $\triangle$ - fixed concentration, ■ - fixed pressure; single shot spectra: ● - fixed pressure). The detection limit for averaged spectra is indicated (dotted).

5.4   Results and discussion

5.4.1   NO depletion in a single 1 ms plasma pulse

In a first series of experiments the evolution of the NO signal during the plasma pulse was studied by changing the delay t_{del} with respect to the ignition of the discharge where zero delay represents a measurement without a discharge. The integrated absorption coefficient K_{NO} was then calculated from the converted raw spectra, corrected with the calibration function and finally normalised to the value without discharge K_{NO} ($t_{del} = 0$). Figure 5.9 is a comparison between measurements performed under flowing and static gas conditions.

At first it turned out that the theoretically achievable time resolution of 90 ns is not necessarily required for the investigation of the plasma kinetics in the present case. Hence the measurements have been taken with a step width between 5 μs and 100 μs. It can be clearly seen from figure 5.9 that K_{NO} is continuously decreasing while the discharge is switched on. For static conditions a noticeable change from the initial integrated absorption coefficient can be observed 200 μs after the breakdown. At the end of the pulse K_{NO} has dropped to 80 % of its initial value. When the valves of the discharge tube are open, i.e. with a constant gas flow during the plasma on-phase, the decline of K_{NO} starts even earlier resulting in systematically lower values compared to those under static gas conditions. The difference between the K_{NO}

evolution in a closed (static) and in an open (flow) reactor is likely due to a depletion of the neutral gas density n under flowing conditions caused by the heating of the gas. For a first estimate the pressure p in the tube is assumed to be about constant in the case of an open (flow) reactor. This is a rough approximation since the sound wave speed is in the range of $350 \, \text{ms}^{-1}$. Thus it takes about 700 µs to travel half the tube length (the reactor is open at both sides). During the plasma pulse the gas is heated up to the same value in static and flowing conditions, convection can be neglected under the present experimental conditions. According to the ideal gas law $p = nk_BT$ for a nearly unchanged pressure the number density of the neutrals including NO is depleted supported by the constant gas flow through the tube whereas this is circumvented when the valves are closed, i.e. under static conditions. Hence the difference in the values of K_{NO} between flowing and static conditions represents the number of depleted NO molecules, as will be discussed in section 5.4.4.

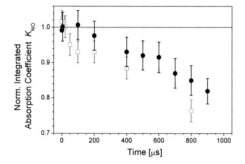

Figure 5.9:
Decay of the normalised integrated absorption coefficient of NO for flowing (□) and static (●) gas conditions for a short plasma pulse in an $Ar/(N_2)/NO$ mixture.

It should be pointed out that the mismatch between the values of K_{NO} for the two cases studied cannot be explained exclusively by the estimated upper limit for the error of the data points (figure 5.9) which is typically about 5 %. The small natural fluctuations in the intensity and in the initial emission frequency among the laser pulses are the main contribution to this considerably high deviation with respect to the overall observed changes in K_{NO}. As a result, when averaging a certain number of spectra, as has been done for the flowing gas conditions, the absorption line becomes blurred caused by an inevitable jitter of the trigger regime. Additionally, the intensity fluctuations are mainly affecting the ratio of I_0/I and consequently the integrated absorption coefficient. Nevertheless, the difference between flowing and static gas conditions systematically exceeds the error induced by the factors mentioned above.

5.4.2   NO depletion with multiple plasma pulses

The evolution of the NO absorption signal in a series of 50 successive plasma pulses is shown in figure 5.10. The signal was monitored 900 µs after the beginning of each pulse. The reactor is closed all the time without renewing the gas. Without any discharge an initial NO number density of $5.8 \times 10^{14} \, \text{cm}^{-3}$ was deduced. The corresponding integrated absorption coefficient was again used to normalise the values of K_{NO} that were measured with the discharge.

Figure 5.10 shows an immediate drop in K_{NO} that occurs during the first pulse. The integrated absorption coefficient is lowered to 81 % which is consistent with the measured decrease after one single pulse under static conditions (figure 5.9). For the following 5 to 8 pulses the integrated absorption coefficient remains constant and starts to decrease slowly thereafter. Finally, after 50 pulses about 25 % of the initial K_{NO} is measured. The significant difference between the effect of the first plasma pulse, leading to a 20 % decay of K_{NO} as indicated in figure 5.10 with the dashed line, and the next plasma pulses suggests that different mechanisms are involved. Some aspects of molecular reactions characterised by different time scales, including the excitation of higher vibrational levels, have been analysed in [13] and more recently in [34].

In order to scrutinise the different reactions involved in the gas phase kinetics longer plasma pulses were applied to a gas mixture containing NO and only one bath gas, namely N_2, where also plasma chemical models exist for [34].

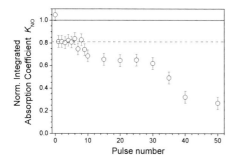

Figure 5.10:
Decay of the normalised integrated absorption coefficient of NO for multi-pulse experiments in an $Ar/(N_2)/NO$ mixture. The dashed line represents the sudden drop in K_{NO} during the first pulse.

5.4.3   NO depletion in a single 100 ms plasma pulse

The decay of the NO signal was studied for both flowing and static gas conditions in single 100 ms long plasma pulses employing currents of 19 mA (figure 5.11) and 35 mA (figure 5.12). In both cases the temporal behaviour of K_{NO} in the on-phase may be described by three time windows showing a characteristic slope of the decrease of NO during the pulse. Similar to short pulses, the decay of K_{NO} during the first 1 ms is within 10 % of the initial value. Depending on the applied current a transition to a strong depletion of the NO signal occurs between 2 and 8 ms. The transition appears thereby earlier at high DC currents. The slope is similar in both cases until the NO signal levels off at 20 % of the initial K_{NO}. The plateau, which is reached after 20 ms and 50 ms for high and low current respectively, corresponds approximately to a number density of 1×10^{14} cm^{-3} and should be considered as NO detection limit in these experiments, hough this is still higher than the theoretical detection limit (section 5.3.2). The increase of K_{NO} under flowing conditions in the off-phase is connected with the residence time of the system (~ 5 s). Comparing flowing and static gas condition for both currents reveals that at low current no difference between the K_{NO} values can be detected within the error bars (figure 5.11). For high currents a NO depletion due to gas heating might be present. However, the difference is less pronounced, especially during the first 1 ms, than observed in the short pulse experiments where Ar was the main bath gas. This suggests a less efficient gas heating in N_2.

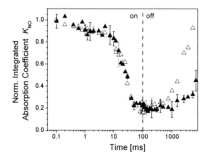
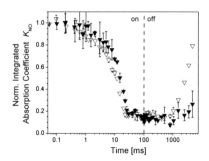

Figure 5.11: Decay of the normalised integrated absorption coefficient of NO for flowing ($\triangle$) and static ($\blacktriangle$) gas conditions for a long plasma pulse in a N_2/NO mixture at 19 mA discharge current. Transition from on- to off-phase is indicated (dashed).

Figure 5.12: Decay of the normalised integrated absorption coefficient of NO for flowing ($\triangledown$) and static ($\blacktriangledown$) gas conditions for a long plasma pulse in a N_2/NO mixture at 35 mA discharge current. Transition from on- to off-phase is indicated (dashed).

5.4.4   Temperature evolution during a plasma pulse

From the results shown particularly in figure 5.9 it was concluded that a part of the decrease of K_{NO} in the case of flowing gas conditions can be attributed to the depletion of the neutrals due to an increase of the gas temperature. Unfortunately, in the present study the line profile of the NO absorption lines is affected by the rapid passage effect and no Doppler width can be inferred. However the measured difference of K_{NO} between an open and a closed reactor may be used to estimate the temperature increase during one single plasma pulse using the ideal gas law. Since the gas temperature variation is identical in the two conditions, we get from equations (2 - 3) and (2 -6):

$$\frac{\Delta K_{NO}}{<K_{NO}>}(t) \approx \frac{\Delta n(NO)}{<n(NO)>}(t) \approx \frac{T(t)-T(0)}{T_{avg}(t)},$$ (5 - 1)

where $<...>$ is the average and $\Delta...$ is the difference respectively between flowing and static conditions. Replacing each number density $n(NO)$ in the second term of equation (5 - 1) by means of the ideal gas law and further simplification lead to the third term of (5 - 1). Here $T(0)$ is the gas temperature at $t = 0$ which is linked with the number density under static conditions fixed all the time and $T(t)$ is the gas temperature at the time t linked with the number density under flowing conditions. Consequently, $T_{avg}(t)$ is the average of $T(0)$ and $T(t)$ and linked with $<n(NO)>$. This provides only a rough relative measure of the temperature variation, which is between 10 % and 15 % after one plasma pulse (figure 5.13) corresponding to an increase of 35 K in terms of absolute values of the temperature (figure 5.14). The error for the experimentally obtained values of the gas temperature is large, especially at the beginning of the pulse, since the difference of two similar values contributes to the quotient in equation (5 - 1) and the pressure cannot be considered as absolutely constant on very short time scales. However, an increase above room temperature can be clearly seen from the discrete absolute values given in figure 5.14.

In order to analyse the evolution of the temperature more precisely a simplified solution of the time dependent heat transfer equation is used [38]. The main ideas of the approach and the adaptation to the present case are summarised in the appendix C.1.

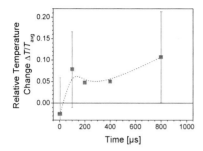

Figure 5.13: Relative temperature change inferred from the integrated absorption coefficients versus time. ΔT is the difference and T_{avg} is the average of $T(0)$ (gas temperature at $t = 0$, linked with $n(\mathrm{NO})$ under static conditions) and $T(t)$ (gas temperature at the time t, linked with $n(\mathrm{NO})$ under flowing conditions) The dashed line is used as a guide to the eye.

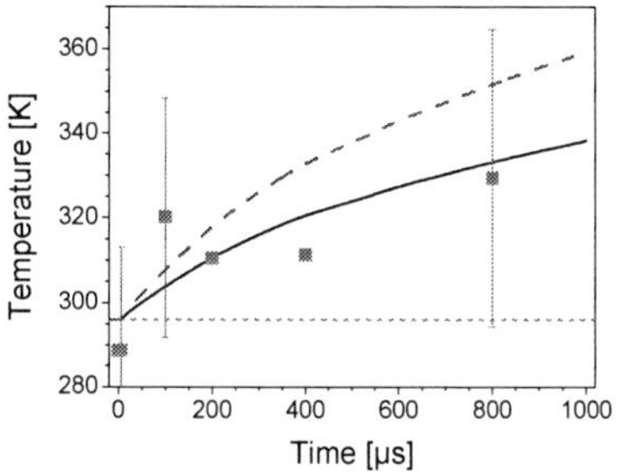

Figure 5.14: Comparison between the gas heating dynamics estimated from the experiments (symbols) and obtained by modelling. Calculations are displayed for the tube centre (dashed) and after spatially averaging over the full tube cross section (solid). The dotted line represents the wall or room temperature respectively.

After simplifying the heat transfer equation the temporal evolution of the gas temperature at the tube axis $T_0(t)$ can be derived numerically by means of equation (C - 5). For this, the set starting condition is $T_0(t) = T_w = 296$ K. Applying equation (C - 4) a spatially averaged temperature over the plasma volume is finally deduced. This value fits properly to the experimental conditions since in practice the laser beam cross section has a finite size which covers more than just the tube axis. Consequently, an average is measured over a part of the radial density profile as well as the temperature profile. For a comparison with the measured values both a spatially averaged and the central temperature are therefore calculated. Since the repetition rate of the discharge is 1 Hz (short pulse) with a pulse on-time in the same order as the time of the gas heating τ_{heat}, i.e. a few milliseconds, it can be assumed that the system has enough time to equilibrate after one pulse. Thus the assumption that the heating starts at room temperature is always valid[3]. In addition to the characteristic values n_m, λ_g, c_p and τ_{heat} for the gas mixture, further input data are mandatory, namely the time dependent current density $j(t)$ and electric field $E(t)$. For both of them the measured values were used.

The result of the temporal evolution of the temperature at the tube centre and the spatially averaged value is shown in figure 5.14. Fairly good agreement is found in comparison to the values obtained from K_{NO} when taking into account the underlying assumptions for the data estimated from K_{NO}, e.g. 800 µs after the ignition of the discharge the calculation yields 333 K (351 K in the centre) compared to 329 K. The calculated gas

---

[3] This holds as well for the 100 ms long plasma pulses because the repetition rate was adapted to 0.1 Hz.

temperature increases continuously, but quite slowly during the pulse. From that, it is also clear that the apparently high temperature at 100 μs, which is inferred from K_{NO}, is more likely due to measurement inaccuracies rather than a real temperature and is particularly not related to the peak power measured at the beginning of the pulse during the breakdown (figure 5.4, lower panel). Notably, the temperatures in the discharge centre and averaged over the entire cross section fall within the experimental error bars and should therefore be considered as upper and lower limit respectively.

The computed averaged temperature including temperature dependent values of n_m and λ_g in equation (C - 6) is 338 K when the discharge is switched off. The (characteristic) heating time of the system varies between 2.1 ms and 1.6 ms from the ignition till the end of the discharge indicating that the dynamics of the gas heating is unaltered. Since n_m and λ_g, that are mainly affected by the temperature, vary by not more than 20 % due to the heating, the error introduced by assuming constant values is almost negligible: the temperature at the end of the plasma pulse would then be 337 K.

The main limitations of such a basic analytical model are i) Only the spatially averaged injected power density $j(t)E(t)$ is considered. Actually, in DC discharges, the power is transferred into the plasma via electrons, whose density decrease from the tube axis to the walls; ii) The power is assumed to be instantaneously transferred to the neutrals as kinetic energy. It is well known that a part of the power is used for ionisation, dissociation of molecules, and electronic and vibrational excitation. In particular, in N_2 containing plasmas, vibrational excitation is an important energy loss process for electrons but vibrational-translational exchange is a slow process [10]. This means that a part of the injected power may be stored in degrees of freedom other than the kinetic energy of the atoms and molecules on time scales of a few hundreds of microseconds. Therefore a precise calculation of the time evolution of the gas temperature would require a detailed analysis of the electron and vibrational kinetics which is far beyond the scope of this thesis. Thus in the following, the calculated temperature is considered as the maximum achievable gas temperature.

In particular, limitation (ii) hinders a reasonable prediction of the gas temperature in the N_2/NO system used for long plasma pulses. A straightforward calculation for the 35 mA discharge suggests a temperature increase up to 510 K after 10 ms at the tube centre (and 440 K after spatially averaging), which appears unrealistic. According to (5 - 1) the difference in K_{NO} between flowing and static conditions in figure 5.12 can be again exploited to estimate the relative and subsequently the absolute gas temperature of the N_2 discharge (see appendix, figure C.1). This suggests an increase of less than 20 % after 10 ms. Nevertheless higher temperatures may be achieved later in the pulse when vibrational-translational exchange occurs. A straightforward downscaling of the injected power density yields that less than 50 % of the injected power are directly or indirectly used for neutral gas heating. This fraction was obviously higher for the Ar discharge (calculations were made with 100 %).

5.4.5   Influence of the temperature on the spectroscopic results

In the previous sections the change of the gas temperature was predicted from the difference between the integrated absorption coefficients obtained for flowing and static gas conditions and proven for discharges with a small N_2 content by a model calculation. Consequently, the question arises how the spectroscopic results are influenced by the increasing temperature.

According to equation (2 - 20) the line strength $S(T)$ is not constant. Therefore the integrated absorption coefficient K_{NO}, comprising $n(NO)$ and $S(T)$, has been used so far instead of calculating number densities of NO. By means of the calculated evolution of the temperature it is now possible to correct the line strength given in the *HITRAN* database for 296 K. For the temperatures obtained from the calculations the factor for the stimulated emission in equation (2 - 20) can be neglected, since its influence is much less than 0.5 %. However, the Boltzmann factor and the partition function vary considerably. The upper panel of figure 5.15 shows the relative change in the line strength compared to the reference value of the database. The influence of the Boltzmann factor and the partition function have been calculated separately yielding an increase of 10.7 % and a decrease of 15.3 % respectively for a temperature of 340 K. In total the line strength drops by 6.5 %. In earlier published results [37] the contribution of the Boltzmann factor was underestimated. The general conclusions are only gradually influenced as will be discussed here.

Although the temperature increase of less than 40 K is relatively moderate, the effect on the line strength is non-negligible (about one third of the total drop). The calculated change in $S(T)$ has been cross-checked experimentally by filling the discharge tube with NO and heating the whole tube externally with heating tape. Without any discharge the absorption of NO has been measured, thereby falling below the calculations for higher temperatures[4] which may attributed to additional thermal conversion inside the tube (bottom of figure 5.15).

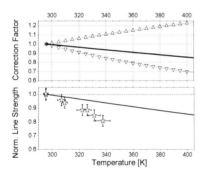

Figure 5.15: **Upper:** Relative temperature dependence of the line strength S (thick solid line) as a result of the temperature dependent Boltzmann term ($\triangle$) and the partition function ($\triangledown$). **Lower:** Experimentally observed temperature dependence ($\star$) of S compared to the calculation (solid line).

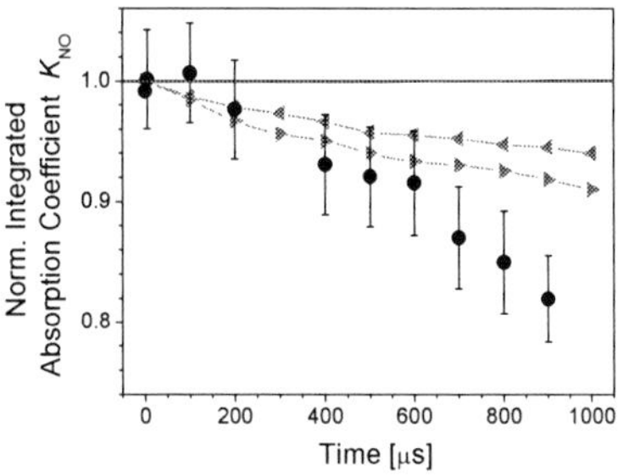

Figure 5.16: Recalculated integrated absorption coefficient of NO for static gas conditions using the modelled temperature increase at the tube centre ($\blacktriangleright$) or spatially averaged over the tube cross section ($\blacktriangleleft$) and its influence on the line strength in comparison with the experiment ($\bullet$).

Actually, the decrease in $S(T)$ as given above contributes significantly to the decrease in K_{NO} that has been observed in the short plasma pulse. By combining the computed evolution of the spatially averaged temperature ($<T(t)>_{vol}$) or the tube centre temperature $T(t)_{centre}$ with $S(T)$ it is now possible to distinguish between the temperature effect on the line

---

[4] Note that in the case of the underestimated Boltzmann factor the experimental results were totally in accord with the calculations [37].

strength and possible chemical effects on the number density n(NO) during the plasma pulse. Both of which may lead to the decrease in K_{NO}. Assuming that the NO density remains constant during a short plasma pulse then all changes in K_{NO} are attributed to changes in the line strength. The consequence for the integrated absorption coefficient for static gas conditions, the data already presented in figure 5.9, is shown in figure 5.16. Using the modelled temperature and line strength values $S(<T(t)>_{vol})$ and $S(T(t)_{centre})$ has been determined for each moment of the entire plasma pulse and multiplied with n(NO) which is assumed to be constant. (The same calculation was carried out for the 35 mA long pulse N_2/NO discharge; the trace of $S(T(t))$ is depicted in the appendix (figure C.2).)

Specifically at the early plasma stage of the short pulse the decrease in the line strength may entirely compensate the decrease in K_{NO}. After 400 µs the observed absorption coefficient is smaller than the predicted one if no further plasma chemical reactions are considered (figure 5.16)[5]. The discrepancy between the calculated and measured drop in K_{NO} up to 600 µs is still within the error bars. However, later in the (short) pulse plasma chemical removal of NO is reasonable.

5.4.6   Heavy species kinetics in pulsed DC discharges containing N_2

Recently, a model of the N_2 - O_2 chemistry for the presently used pulsed DC reactor has been developed [34]. The comparison between experimentally observed and computed NO number densities in synthetic air yielded fair agreement. For this reason, the same model has been applied to the present long pulse studies using a N_2/NO gas mixture where clearly less O_2 is available than in the originally studied air plasmas. The aim was to identify main reactions leading to the two different slopes in the temporal NO behaviour. The main reactions (a) - (e) contributing to the depopulation and population of the NO(X) ground state are plotted in figure 5.17 for the pulsed discharges employing 19 mA and 35 mA DC current respectively.

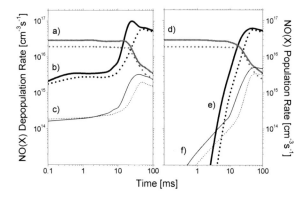

Figure 5.17:   NO(X) depopulation (left, a - c) and population (right, d - e) rates obtained from a time dependent plasma chemical model [34] for two different currents (19 mA - dotted, 35 mA - solid). For reactions a) - e) see text.

---

[5]   By underestimating the Boltzmann factor it was earlier concluded that there is entirely no plasma chemical removal of NO in short pulses since the decrease in $S(T)$ overcompensated the decrease in K_{NO} [37].

$$NO + N_2(A) \quad \rightarrow \quad NO(A) + N_2 \quad\quad\quad (a)$$
$$NO + N \quad\quad \rightarrow \quad N_2(X,v) + O \quad\quad (b)$$
$$NO + N(^2D) \quad \rightarrow \quad N_2 + O \quad\quad\quad\quad (c)$$

$$NO(A) \quad\quad\quad \leftrightarrow \quad NO(X) + hf \quad\quad (d)$$
$$N_2(X,v) + O \quad \leftrightarrow \quad NO + N \quad\quad\quad (e)$$
$$N_2(A) + O \quad\quad \leftrightarrow \quad NO + N(^2D) \quad\quad (f)$$

The main depopulation of $NO(X)$ occurs via excitation by the N_2 metastable state (a) showing the highest rate and N atom impact (b, c). Population of $NO(X)$ is accomplished by radiative deexcitation (d) and reactions with vibrationally excited or metastable N_2 molecules (e, f). It transpires, however, that those reactions with the highest rate (a, d) compensate each other. Consequently, during the first 10 ms of the plasma pulse reaction b) and c) have no back reaction of equivalent rate (b vs. e) and (c vs. f), since a significant population of higher vibrationally excited states is reached after ~ 10 ms [34]. It is evident from figure 5.17 that the strong depletion of NO between 5 and 10 ms is caused by an increase of both depopulation rates of N atom based reactions by one order of magnitude.

The mixing ratios obtained from the model calculations accord qualitatively very well with experimental observations within the error bars (figure 5.18). Additionally, the experimental trend that the depletion of NO starts earlier in the pulse at higher currents is confirmed. The obvious discrepancy between calculations and measurements after 20 ms is connected to the detection limit of the method. Experimental data for 35 mA in figure 5.18 have been corrected for $S(T)$ (cf. figure C.2). Since the measurements under static and flowing conditions for 19 mA reveal almost no difference, it can be concluded that the temperature increase as well as a correction for $S(T)$ is negligible and would not lead to a substantially better agreement.

The slight mismatch between model calculations and experimental data in the 5 ... 20 ms range might be already related to surface reactions. While for the short pulse regime surface reactions were too slow, the ms range should be sufficient for a significant surface coverage with plasma produced radicals. As discussed in chapter 3, plasma chemical systems in excess of N_2 tend to form N_2O rather than NO. In this case any surface would dominantly be covered by N atoms and add an additional loss channel for NO.

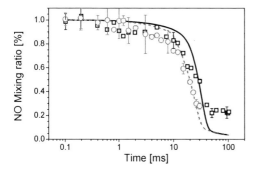

Figure 5.18:
Comparison of NO mixing ratios between (static) experimental data (symbols) and model calculations (lines). The N_2/NO discharge was operated with 19 mA ($\square$, solid) and 35 mA ($\bigcirc$, dashed) at 2.66 mbar and 100 ms pulse length. Experimental data for 35 mA are corrected for the $S(T)$ influence (negligible for 19 mA).

5.5   Summary and conclusions

QCLAS measurements in the mid infrared have been performed on pulsed DC discharges containing a small amount of NO in an Ar - N_2 mixture and in a pure N_2 background. The decay of NO has been followed on the µs and sub-µs timescale using the *intra* pulse mode of operation for the QCL. The minimal time resolution of 90 ns is only limited by the pulse width of the QCL. A calibration procedure was applied to avoid misleading results due to the perturbed line shape of the observed NO line. Experiments have been performed for static and flowing gas conditions during a single plasma pulse exhibiting both an apparent decrease of the NO concentration. From the difference between the two conditions the increase of the gas temperature could be deduced. The dynamics of the gas heating was then refined by a simple model calculation on the basis of the heat transfer equation adapted to the actual discharge setup. The increase of the temperature in the Ar containing plasma during the pulse is not greater than 40 K, but has a significant influence on the line strength of the NO line used. The agreement between model calculations and experiments is better for discharges with less content of N_2. In pure N_2 discharges a fraction of up to 50 % of the injected power was estimated to be transferred into neutral gas heating being more pronounced after ~ 10 ms when higher vibrational levels become populated.

At least one third of the 20 % decrease in the integrated absorption coefficient that has been observed in short pulse plasmas is just due to the heating of the gas which in turn makes the line strength vary in the same order with time and gives an apparent drop of NO. The QCLAS measurements accompanied by the model calculations are a powerful non-invasive temperature probe with a remarkable time resolution up to the sub-µs time scale.

A comparison between calculations of a recently developed model for N_2 - O_2 pulsed discharges and the time resolved QCLAS observations revealed that the loss of NO via $N_2(A)$ is generally compensated, i.e. the depletion in the early pulse phase is dominated by uncompensated N atom reactions which become very efficient after ~ 5 ms. An increasing mismatch between computations and experimental results observed beyond 5 ms require further effort along these lines.

C Appendix

C.1 Temperature evolution in a DC discharge tube

In what follows the adaption of the time dependent heat transfer equation to the present geometry and main aspects are briefly discussed [38].

According to the geometry of the tube configuration the heat transfer equation is written in radial coordinates:

$$n_m \, c_p \, \frac{dT(r,t)}{dt} - \lambda_g \nabla^2 T(r,t) = \frac{P(t)}{\pi \, \rho^2 \, l} = j(t) \cdot E(t), \qquad (C-1)$$

where n_m is the molar density of the gas, c_p is the molar heat capacitance at constant pressure, λ_g is the thermal conductivity, ρ and l are the tube radius and length respectively, j is the current density and E is the electric field strength of the discharge. Although (C - 1) requires numerical computation, its complexity can be reduced by some simplifying assumptions. As shown below it is still possible to reach an adequate description of the problem. In order to obtain a differential equation depending only upon time, equation (C - 1) can be averaged over the plasma volume:

$$\frac{1}{\pi \, \rho^2 \, l} \iiint_{volume} \left(n_m c_p \, \frac{dT(r,t)}{dt} - \lambda_g \nabla^2 T(r,t) \right) r dr d\theta dz = j(t) \cdot E(t). \qquad (C-2)$$

Provided that $T(r,t)$ is only spatially dependent on r, the first term in the integral becomes $n_m c_p (d<T>_{vol}/dt)$. Assuming further a quadratic decrease of $T(r)$ across the plasma radius, as in the steady state, the expression for $T(r)$ is:

$$T(r,t) = T_0(t) - \left(T_0(t) - T_w\right)\left(\frac{r}{\rho}\right)^2, \qquad (C-3)$$

$T_0(t)$ and T_w being the gas temperature at the tube axis and at the wall respectively. T_w is considered to be constant during the plasma pulse and set to 296 K in what follows. After replacing the averaged temperature over the plasma cross section $<T>_{vol}$ by its value as a function of T_0 and T_w by means of an integration of (C - 3) over the radius r from (0 ... ρ), which leads to

$$<T>_{vol} = \frac{2}{3} T_0 + \frac{1}{3} T_w, \qquad (C-4)$$

equation (C - 2) reduces finally to

$$\frac{dT_0}{dt} + \frac{T_0}{\tau} = \frac{1}{\tau}\left(\frac{j(t) \cdot E(t) \cdot \rho^2}{4\lambda_g} + T_w \right), \qquad (C-5)$$

where τ_{heat} is

$$\tau_{\text{heat}} = \frac{n_m \rho^2 c_p}{6\lambda_g}, \tag{C - 6}$$

which has the form of a characteristic heating time for the system. Hence n_m, λ_g and c_p depend on the temperature with n_m being proportional to T^{-1} and λ_g is an increasing function of T, whereas c_p varies only slowly with T. Thus (C - 5) is a non-linear equation that was solved numerically.

However at first, further simplifications are made to estimate the value of τ. In the present case, the experimental temperature increase shown in figure 5.13 is less than 15 %, so that in a first approximation, constant values for n_m, λ_g and c_p are assumed. In the first mixture used consisting of nearly 10 % nitrogen in argon, the heat capacitance at 300 K is taken as $c_p(\text{mixture}) = 0.1 \cdot c_p(N_2) + 0.9 \cdot c_p(Ar) = 21.6$ J mol^{-1} K^{-1} [39]. For the thermal conductivity the analogous formula is applied resulting in $\lambda_g(\text{mixture}) = 0.0185$ W m^{-1} K^{-1} [40]. For the experimental conditions (296 K, 2.66 mbar) the molar density is $n_m = 1.07 \cdot 10^{-2}$ mole m^{-3}. The calculated τ is then of 2.1 ms which is about two orders of magnitude higher than the time for the breakdown process and bigger than the entire (short) plasma pulse indicating that the heating is a slow process in respect to the short plasma pulse. When taking into account the temperature dependence of the values contributing to it, τ can obviously no longer be referred to as a constant "characteristic time" since it varies (slowly) with T as well (typically between 2.1 and 1.5 ms for the Ar/N$_2$ mixture).

C.2   Supplemental figures for long plasma pulses

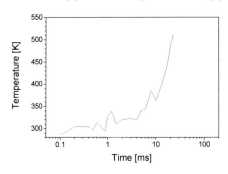

Figure C.1:
Estimated increase of the absolute gas temperature inferred from the relative difference between the integrated absorption coefficients of NO, K_{NO}, under static and flowing conditions (eq. (5 - 1)). Experimental data were taken and interpolated from figure 5.12 for a 100 ms long N$_2$/NO discharge (35 mA). Data close to the NO detection limit for $t > 20$ ms were considered.

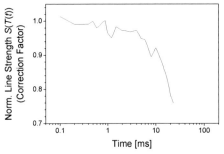

Figure C.2:
Correction factor for the extracted integrated absorption coefficients of NO measured in a 100 ms long N$_2$/NO discharge (35 mA). The correction is necessary due to the temperature dependence of the line strength. $S(t)$ was deduced by means of figure C.1 (i.e. $S(T(t))$).

Bibliography

[1]   E.M. van Veldhuizen *Electrical Discharges for Environmental Purposes. Fundamentals and Applications*, ISBN 978-156072743-9, (NOVA Science Publishers, Hauppauge, 2000).

[2]   *Non-Thermal Plasma Techniques for Pollution control*, eds. B. M. Penetrante, S. E. Schultheis, ISBN 0-387-57174-4, (Springer, New York, 1993).

[3]   B.M. Penetrante, M.C. Hsiao, J.N. Bardsley, B.T. Merritt, G.E. Vogtlin, A. Kuthi, C.P. Burkhart, J.R. Bayless, *Plasma Sources Sci. Technol.* **6**, 251 (1997).

[4]   T. Hammer, *Plasma Sources Sci. Technol.* **11**, A196 (2002).

[5]   T. Yamamoto, *J. Electrostatics* **42**, 227 (1997).

[6]   E. Filimonova, R. Amirov, H. Kim, I. Park, *J. Phys. D: Appl. Phys.* **33**, 1716 (2000).

[7]   J. Mc Adams, *J. Phys. D: Appl. Phys.* **34**, 2810 (2001).

[8]   F. Fresnet, G. Baravian, L. Magne, S. Pasquiers, C. Postel, V. Puech, A. Rousseau, *Appl. Phys. Lett.* **77**, 4118 (2000).

[9]   K. Yan, S. Kanazawa, T. Ohkubo, Y. Nomoto, *Plasma Chem. Plasma Process.* **19**, 421 (1999).

[10]   M. Capitelli, C.M. Ferreira, B.F. Gordiets, A.I. Osipov, *Plasma Kinetics in Atmospheric Gases*, ISBN 3-540-67416-0, (Springer, Berlin, 2000).

[11]   A. Rousseau, A. Dantier, L.V. Gatilova, Y. Ionikh, J. Röpcke, Y.A. Tolmachev, *Plasma Sources Sci. Technol.* **14**, 70 (2005).

[12]   Y. Ionikh, A.V. Meshchanov, J. Röpcke, A. Rousseau, *Chem. Phys.* **322**, 411 (2006).

[13]   L.V. Gatilova, K. Allegraud, J. Guillon, Y.Z. Ionikh, G. Cartry, J. Röpcke, A. Rousseau, *Plasma Sources Sci. Technol.* **16**, S107 (2007).

[14]   C.D. Pintassilgo, J. Loureiro, V. Guerra, *J. Phys. D: Appl. Phys.* **38**, 417 (2005).

[15]   K. Kutasi, C.D. Pintassilgo, P.J. Coelho, J. Loureiro, *J. Phys. D: Appl. Phys.* **39**, 3978 (2006).

[16]   B.F. Gordiets, C.M. Ferreira, V.L. Guerra, J.M.A.H. Loureiro, J. Nahorny, D. Pagnon, M. Touzeau, M. Vialle, *IEEE Trans. Plasma Science* **23**, 750 (1995).

[17]   V. Guerra, J. Loureiro, *J. Phys. D: Appl. Phys.* **28**, 1903 (1995).

[18]   V. Guerra, J. Loureiro, *Plasma Sources Sci. Technol.* **6**, 373 (1997).

[19]   J. Nahorny, C.M. Ferreira, B. Gordiets, D. Pagnon, M. Touzeau, M. Vialle, *J. Phys. D: Appl. Phys.* **28**, 738 (1995).

[20]   C.D. Pintassilgo, K. Kutasi, J. Loureiro, *Plasma Sources Sci. Technol.* **16**, S115 (2007).

[21]   G. Oinuma, Y. Inanaga, S. Noda, Y. Tanimura, M. Kuzumoto, Y. Tabata, K. Watanabe. *J. Phys. D: Appl. Phys.* **41**, 155204 (2008).

[22]   B. Gordiets, C.M. Ferreira, J. Nahorny, D. Pagnon, M. Touzeau, M. Vialle, J. Phys. D: Appl. Phys. 29, 1021 (1996).

[23]   J.H. van Helden, R.A.B. Zijlmans, D.C. Schram. R. Engeln, *Plasma Sources Sci. Technol.* **18**, 025020 (2009).

[24]   R.A.B. Zijlmans, Ph.D. Thesis, Eindhoven University of Technology, ISBN 978-90-386-1288-1, 2008.

[25]   J.B. McManus, D. Nelson, M. Zahniser, L. Mechold, M. Osiac, J. Röpcke, A. Rousseau, *Rev. Sci. Instr.* **74**, 2709 (2003).

[26]   B.M. Gorshunov, A.P. Shotov, I.I. Zasavitsky, V.G. Koloshnikov, Y.A. Kuritsyn, G.V. Vedeneeva, *Opt. Comm.* **28**, 64 (1979).

[27]   D.D. Nelson, J.B. McManus, S. Urbanski, S. Herndon, M.S. Zahniser, *Spectroch. Acta* Part *A* **60**, 3325 (2004).

[28]   M.L. Silva, D.M. Sonnenfroh, D.I. Rosen, M.G. Allen, A. O'Keefe, *Appl. Phys. B* **81**, 705 (2005).

[29]   O. Gabriel, S. Stepanov, M. Pfafferott, J. Meichsner, *Plasma Sources Sci. Technol.* **15**, 858 (2006).

[30]   S. De Benedictis, G. Dilecce, M. Simek, *J. Phys. D: Appl. Phys.* **30**, 2887 (1997).

[31]   S. De Benedictis, G. Dilecce, *J. Phys.* III **6**, 1189 (1996).

[32]   M. Castillo, V.J. Herrero, I. Mendez, I. Tanarro, *Plasma Sources Sci. Technol.* **13**, 343 (2004).

[33]   G. Cartry, L. Magne, G. Cernogora, *J. Phys. D: Appl. Phys.* **32**, 1894 (1999).

[34]   C.D. Pintassilgo, O. Guaitella, A. Rousseau, *Plasma Sources Sci. Technol.* **18**, 025005 (2009).

[35]   L.S. Rothman *et al.*, *J. Quant. Spectrosc. Radiat. Transfer* **96**, 139 (2005).

[36]   Y.P. Raizer, *Gas Discharge Physics*, ISBN 3-540-19462-2, (Springer, Berlin, 1991).

[37]   S. Welzel, L. Gatilova, J. Röpcke, A. Rousseau, *Plasma Sources Sci. Technol.* **16**, 822 (2007).

[38]   A. Rousseau, E. Teboul, N. Sadeghi, *Plasma Sources Sci. Technol.* **13**, 166 (2004).

[39]   B.J. McBride, S. Gordon, M.A. Reno, *NASA Technical Paper* **3287** (1993).

[40]   C.Y. Ho, R.W. Powell, P.E. Liley, *J. Phys. Chem. Ref. Data* **1**, 279 (1972).

6   Trace gas measurements using optically resonant cavities and quantum cascade lasers

6.1   Introduction

The development of highly sensitive, compact and robust optical sensors for molecular detection is of interest for an increasing number of chemical sensing applications[1], such as environmental monitoring and atmospheric chemistry [1,2], plasma diagnostics [3,4], combustion studies [5], detection of explosives [6] and medical diagnostics [7,8]. Absorption spectroscopy in the mid infrared (MIR) spectral region using lasers as radiation sources is an effective method for monitoring molecular species. In principle path lengths up to several kilometres can be achieved by using optical resonators for cavity ring down spectroscopy (CRDS), cavity enhanced absorption spectroscopy (CEAS), or integrated cavity output spectroscopy (ICOS). A wealth of studies using optical cavities have been published since the introduction of CRDS by O'Keefe and Deacon in 1988 [9], of CEAS by Engeln et al. in 1998 [10] and of ICOS by O'Keefe et al. in the same year [11,12]. In CRDS the decay of light leaking out of a resonant optical cavity and provided either by a pulsed or interrupted continuous wave (cw) source is monitored in the time domain. Cavity leak-out spectroscopy (CALOS), based on cw excitation of the cavity, essentially operates on the same principle and was introduced by Mürtz et al. in 1999 [13]. In contrast, in CEAS or ICOS the steady-state transmission or integrated transmitted intensity through the cavity is observed as a function of the laser frequency. The most sensitive form so far reported is noise-immune cavity enhanced optical heterodyne molecular spectroscopy (NICE-OHMS) combining frequency modulation and mode locking techniques with optical cavities [14].

The majority of cavity based methods have used sources of radiation in the ultraviolet and visible regions. For many years the infrared spectral range could not be employed either for CRDS or for the CEAS or ICOS techniques, because of the lack of suitable radiation sources with the required power and tuneability, but this situation has now changed. Near infrared (NIR) applications have profited from developments in telecommunications where cheap and compact light sources became available [15,16] in the 1990s whereas similar lasers were not available in the mid infrared range. Early experiments in the latter region were carried out with optical parametric oscillators (OPO) pumped with adequately powerful lasers, e.g. in 1995 by Scherer et al. [17-19]. The application of Raman cells or shifters was another method of choice in several studies of free radicals and stable species among them OH, CH_3, or clusters of water and CH_3I [20-24]. Peeters et al. measured ethylene at 10 μm with a waveguide CO_2 laser in a CEAS configuration [25]. Mürtz et al. performed CALOS experiments with CO_2 and CO lasers to detect C_2H_4, $^{13}CH_4$, H_2CO and OCS respectively [13,26-28]. Continuous progress in non-linear frequency conversion techniques in respect to MIR output power levels and available pump sources (see chapter 2) has renewed the interest in combining OPOs or difference frequency generators (DFGs) with optical cavities for

---

[1]   Additional applications and more examples in respect to chemical sensing that are not purely based on resonant optical cavities are collected in chapter 4. The reviews cited in this introduction already focus on the application of optical cavities.

measuring specifically hydrocarbons (CH_4, C_2H_6) in the 3 µm range [29,30]. Recently, difference frequency conversion based on orientation patterned GaAs enabled CRDS experiments on N_2O in the 7 ... 9 µm range [31] to be carried out. Since this new kind of DFG source is still under development the final sensitivity suffered from the low output power (< 10 µW). In all these cases sophisticated optical geometries were developed which were more suitable for the research laboratory than for field applications. The availability of a research facility was also essential for the free electron laser experiments in the MIR by Engeln et al. who was able to detect C_2H_4 (gas phase) and C_{60} (solid) [32,33].

To overcome the drawbacks of the bulky radiation sources mentioned above, small semiconductor based lasers became increasingly used. In 2001 a MIR cavity ring down spectrometer, using lead salt lasers, and the detection of CO was reported [34,35]. More recently, an attempt of pulsed slit jet CRDS, using a 3 µm lead salt diode laser, has been published [36]. In both studies the technique suffered from low laser intensity and the beam quality.

Recent advances in semiconductor laser technology, in particular the advent of intersubband quantum cascade lasers (QCL) and interband cascade lasers (ICL), provides new possibilities for highly sensitive and selective trace gas detection using MIR absorption spectroscopy [37-45] and enabled sensitivities of 5×10^{-10} $cm^{-1}Hz^{-1/2}$ to be accomplished but at the expense of large sample volumes [40]. Distributed feedback (DFB)-QCLs combine single frequency operation with tuneability over several wavenumbers, and average powers over a mW; hence they are superior to lead salt lasers. While pulsed QCLs working at room temperature have been commercially available for several years, room temperature cw QCLs have only recently been introduced. The relatively high output power of the QCL permits the use of optical cavities with high finesse. With the help of such cavities the effective path length of the laser beam in the absorbing medium can essentially be increased to more than the 200 m limit usually available from conventional optical multi-pass cells [39,46] while keeping the sample and pumped volume small.

The potential of such a combination was demonstrated in CRDS measurements with a cryogenically cooled cw DFB-QCL [47]. A sensitivity of 9.7×10^{-11} $cm^{-1}Hz^{-1/2}$, i.e. within an order of magnitude of the ultimate shot-noise-limit, was achieved with the application of QCLs to NICE-OHMS [48]. Focusing on potential field deployable systems, Silva et al. reported a study which combined a TE cooled pulsed QCL with CRDS and ICOS [49]. A calibration of non-linear experimental results was necessary and the observation of asymmetric line shapes caused by the frequency chirp in a single laser pulse was discussed. Other studies combining TE cooled pulsed QCLs with the CRDS approach recognised the influence of the frequency chirp of QCLs leading to disturbed line profiles followed by reduced sensitivity and selectivity [50,51]. Further development was forthcoming from the availability of DFB-QCLs with increasingly good performance such as output power and near ambient temperature operation, in the beginning limited to the 5 - 10 µm range. Several applications of QCLs combined with either the CRDS or the CEAS/ICOS techniques for trace gas detection [52-59] followed and sensitivities down to 2×10^{-9} $cm^{-1}Hz^{-1/2}$ were reported [53]. Recently, the availability of ICL, which emit in the 3 - 5 µm range, also enabled ICOS experiments on H_2CO to be performed [60].

However, the broader use of cavity based techniques in the MIR in industry or for field application has so far not only been hindered by space or weight restrictions, but also by the need for cryogenic cooling of either the laser or (at least) the detector. Although continuous improvement of the spectrometer design enabled a sensitivity of 9×10^{-10} cm^{-1}Hz$^{-1/2}$ in off-axis configuration to be obtained, the entire system was still based on LN cooled devices [61]. Recently there has been a trend away from liquid nitrogen (LN) cooled systems towards thermoelectrically (TE) cooled devices [56,57], but the performance of an entirely TE cooled system based on optical cavities has not yet been investigated.

TE cooled pulsed and cw QCLs have therefore been employed for high finesse cavity absorption spectroscopy in the MIR. The approach was threefold: (i) CRDS and ICOS experiments have been performed with pulsed QCLs, (ii) a room temperature cw QCL was used in a complementary CEAS configuration aimed at achieving a sensitive, LN free spectrometer for sampling small volumes, and (iii) the spectrometer design was limited to an unstabilised, unlocked cavity to facilitate a straightforward transfer of the optical table arrangement to a planar microwave reactor in the future.

In the CRDS experiment the influence of the frequency chirp of pulsed QCLs on the sensitivity and accuracy of concentration measurements has been analysed in detail using CH$_4$ and N$_2$O as absorbing gases. Apart from several studies on NO at 5 µm the MIR spectral region in general and the 7 - 8 µm region presently used have not been extensively investigated with techniques based on optical resonators.

After a short compilation of the necessary formulae for data analysis (section 6.2), the experimental arrangements for the different approaches are presented in section 6.3. Sections 6.4.1 and 6.4.2 follow with a detailed analysis of the CRDS and ICOS experiments with the pulsed QCLs. The CEAS system is validated in section 6.4.3. Finally, the conclusions are presented in section 6.5.

6.2 Theoretical considerations

This section summarises the basic equations which are necessary to analyse the recorded data. A more thorough introduction with a detailed discussion of different approaches and the assumptions lying behind the formulae which are in common use was recently published in a comprehensive review by Mazurenka et al. [62]. The fundamental equation for a laser absorption experiment is the Beer-Lambert law (eq. (2 - 4)) where L_{eff} is the effective absorption path length in the absorbing medium, and depends on which method is applied.

Following the description of Mazurenka et al. only a few restrictions are used for the ballistic assumption (or ping-pong model) to describe the behaviour of a laser pulse shorter than the round trip time in an optical resonator, namely that the reflectivity of the cavity mirrors is equal ($R_1 = R_2 \equiv R$) and scattering is neglected, i.e. the transmission $T \approx 1 - R$. The separation of the highly reflective mirrors is L, whereas the interaction length of the laser beam with the absorbing medium is d. The intensity I^0 directly transmitted through the cavity without any additional round trip is then

$$I^0 = (I_{in}\eta T) \cdot e^{-kd} \cdot T,$$ (6 - 1)

where $\eta \, (\leq 1)$ factors in the possibility of non-ideal mode matching of the incident radiation I_{in} with the cavity modes [62]. The definition of the frequency dependent absorption coefficient $k(\nu)$ and its connection to the absorption cross section $\sigma(\nu)$ follow from equation (2 - 3).

After m round trips the intensity leaking out of the cavity I_m is

$$I^m = I^0 \cdot \left(Re^{-kd} \right)^{2m}.$$ (6 - 2)

Although strictly speaking the ballistic assumption is only valid for short laser pulses exponential decays, according to the above mentioned treatment, can be also observed for light pulses longer than the round trip time as long as the absorption line is broader than the spacing of the evolving cavity modes [63].

6.2.1   Cavity ring down effect

Equation (6 - 2) can readily be transformed into the well-known relationship for the cavity ring down intensity

$$I^m = I^0 \cdot \exp\left(-\frac{t_m}{\tau} \right),$$ (6 - 3)

where the decay time τ is defined by

$$\tau(\nu) = \frac{L}{c(k(\nu)d - \ln R)},$$ (6 - 4)

and the time t_m necessary for a cavity round trip is given by $t_m = (2mL)/c$, where c is the velocity of light. In the case of typically high values for the mirror reflectivity, $R \rightarrow 1$, equation (6 - 4) can be approximated by:

$$\tau(\nu) \approx \frac{L}{c((1 - R) + k(\nu)d)}.$$ (6 - 5)

For an empty cavity ($k = 0$) and with $d = L$ (absorbing medium completely filling the cavity), an effective absorption path for the experiment $L_{eff} = c\tau_0 = L/(1 - R)$ can be defined. Transforming (6 - 5) yields the absorption coefficient:

$$n\sigma(\nu) = k(\nu) = \left(\frac{1}{\tau(\nu)} - \frac{1}{\tau_0(\nu)} \right) \frac{L}{dc},$$ (6 - 6)

which can be calculated from measurements in the time domain.

6.2.2 Cavity enhanced absorption

The cw excitation of a cavity (CEAS) was originally introduced by Engeln et al. [10]. The description of the temporal development of the intensity inside the cavity I_{cav} is similar to (6 - 2). The only difference is an additional source term taking into account the continuously available radiation rather than a short laser pulse [53]

$$\frac{d I_{cav}}{dt} = \frac{c}{2L}\left(I_{in}\eta T - 2I_{cav}\left(1 - Re^{-kd}\right)\right),$$ (6 - 7)

leading to a steady state output of the cavity characterised by

$$I_{out}(t) = \frac{I_{in}\eta T^2}{2(1 - Re^{-kd})}\left(1 - \exp\left(-\frac{t}{\tau}\right)\right).$$ (6 - 8)

This is governed both in time and amplitude by the cavity losses (R and k). Hence the laser frequency sweep has to stay long enough in resonance with the cavity modes to enable a sufficient intensity build-up inside the cavity. At the same time the scan rate must be significantly higher than the jittering of the cavity modes to avoid large intensity fluctuations [10]. Using the steady state output the ratio of the intensities for an empty (I_0) and a filled (I) cavity respectively can be defined as

$$\frac{I_0}{I} = \frac{I_{out}(k = 0)}{I_{out}} = 1 + GU,$$ (6 - 9)

where $G = R/(1 - R)$ and $U = (1 - \exp(-kd))$ [53]. Equation (6 - 9) is valid for all R and k. It is easy to show that in the weak absorption limit ($k \to 0$ and $R \to 1$) equation (6 - 9) becomes

$$n\sigma(\nu) = k(\nu) = \left(\frac{I_0(\nu)}{I(\nu)} - 1\right)\frac{1 - R}{d}.$$ (6 - 10)

By analogy with the Beer-Lambert law (2 - 4) assuming weak absorption the effective path length can thus be expressed as $L_{eff} = L/(1 - R)$ for $d = L$.

Equation (6 - 10) can be analysed further. If the natural logarithm is taken from (6 - 9) and plotted against the molecular number density n a linear relationship is expected, as long as a weak absorber is present in the cavity. In this case the expressions can be further simplified:

$$\ln\left(\frac{I_0}{I}\right) \approx GU \sim n.$$ (6 - 11)

Thus equation (6 - 11) fulfils two functions. Firstly, it serves as a check on the validity of the weak absorption limit. Secondly, in the range where $\ln(I_0/I)$ is proportional to n, the slope of this linear relationship enables the mirror reflectivity R to be determined from a known concentration standard and hence provides a means for the absolute calibration of CEAS (or ICOS, as will be shown in section 6.2.3).

6.2.3   Integrated cavity output

ICOS was introduced for pulsed excitation of the cavity [12]. For data analysis there is in fact no difference from CEAS since equation (6 - 10) holds for cw as well as for pulsed excitation. In the latter case the source term in (6 - 7) is absent after the initial excitation. On the other hand, following the ping-pong model, the integrated cavity output I_{out} corresponds to an infinite sum over the intensity I^m leaking out of the cavity.

$$I_{out} = I_{in}\eta(1-R)^2 e^{-kd} \sum_{m=0}^{\infty} \left(Re^{-kd}\right)^{2m} = I_{in}\eta(1-R)^2 \frac{e^{-kd}}{1-\left(Re^{-kd}\right)^2} . \qquad (6 - 12)$$

The sum over m converges for the experimentally reasonable cases when $R \cdot \exp(-kd) < 1$ and yields the right hand side of (6 - 12). The ratio of the time integrated signal of the cavity without ($k = 0$) and with the absorbing medium is correspondingly

$$\frac{I_0}{I} = \frac{I_{out}(k=0)}{I_{out}} = \frac{1-\left(Re^{-kd}\right)^2}{\left(1-R^2\right)e^{-kd}} . \qquad (6 - 13)$$

In the limits of weak absorber ($k \to 0$) and high reflectivity ($R \to 1$) equation (6 - 13) can be simplified to

$$\frac{I_0}{I} \approx 1 + \frac{kd}{1-R} , \qquad (6 - 14)$$

which is the standard formula used for ICOS, leading essentially to the same result as in the CEAS case embodied in equation (6 - 10).

In what follows the terms ICOS and CEAS are thus used synonymously, in other words ICOS is used in the context of pulsed excitation and CEAS in the context of cw cavity excitation. In both cases equation (6 - 10) will be applied to extract the absorption coefficient whereas equation (6 - 6) is applied to the CRDS measurements.

6.2.4   Detection limits

For CRDS the uncertainty in the absorption coefficient Δk can be derived from equation (6 - 6) with $\tau \to \tau_0$

$$\Delta k = \frac{\Delta \tau}{\tau_o^2} \frac{L}{dc} . \qquad (6 - 15)$$

Conventionally the standard deviation of the ring down time is used as a measure of $\Delta \tau$. For CEAS and ICOS an analogous relationship can be found from (6 - 10) with $I \to I_0$

$$\Delta k = \frac{\Delta I}{I_0} \frac{1-R}{d},$$ (6 - 16)

where ΔI represents the intensity fluctuations, either inherent to the light source or induced by mode fluctuations. In principle, the error in the absorption coefficient is even higher since the reflectivity of the mirrors R is not known, i.e. a calibration is required leading to a $\sqrt{2}$ times higher uncertainty in k [62]. In general, another quantity, namely the noise equivalent absorption (NEA), defined in [17,61,62] is used to estimate the detection limit

$$NEA = \Delta k \sqrt{\frac{2}{f}},$$ (6 - 17)

where f is the repetition rate of the measurements, i.e. the number of averaged scans in a 1 s interval. It should be noted that the NEA is not defined consistently throughout the literature. Usually employed expressions are summarised and discussed in [61]; the NEA defined here in (6 - 17) would correspond to the "per scan" category in [61].

6.3   Experimental

Two different kinds of experiment are described here for both of which essentially the same optical setup was used: firstly conventional CRDS with a pulsed light source (section 6.3.1) and secondly CEAS using a cw QCL in order to obtain a highly sensitive LN free sensor scheme in the mid infrared (section 6.3.2). The changes required for the cw source and the different analysis of the data will be outlined in section 6.3.2. Basically, for both approaches the laser frequency was swept by a continuous current ramp. In contrast to locked cavity studies, the cavity length here was not actively changed or dithered nor was the cavity locked to the illuminating light source.

6.3.1   Pulsed cavity excitation

The experimental arrangement using pulsed QCLs is straightforward as shown in figure 6.1. The stable resonator was formed by two high reflectivity mirrors (Los Gatos Research) of diameter 25.4 mm and 1 m radius of curvature. The mirrors also served to enclose the vacuum vessel which was made of standard vacuum components. Hence the vacuum cell determined the mirror separation of 0.432 m. Beam shaping optics (BSO) collected the strongly divergent radiation from the QCL firstly with an off-axis parabolic mirror (OAP) of 49.5 mm diameter (f/0.99) and then with a telescope formed by two additional OAPs which reduced the beam diameter by a ratio of almost 1:5. This was sufficient to effectively match the 20 mm aperture of the cavity. No additional optics were added to suppress possible optical feedback to the laser, since feedback was assumed to be negligible for the current DFB-QCL devices without specific antireflection coating. The radiation leaking out of the cavity was collected by an OAP of 25.4 mm diameter (f/0.64) and directed to a detector and preamplifier with a response time of approximately 70 ns. For CRDS relative calibration was provided by recording the fringes of a Germanium etalon in a reference channel and using interpolation. Absolute calibration was achieved by comparison with standard gas absorption spectra recorded in the reference channel simultaneously with the CRDS spectra. Typically the same stable molecule was used for calibration and cavity experiments. Both signal and reference channel were equipped with LN cooled detectors (Judson Technologies, J15D12 series).

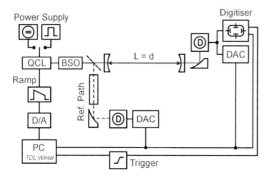

Figure 6.1: Schematic diagram of the apparatus used for both CRDS and CEAS experiments. Beam shaping optics (BSO) provided efficient light transmission to the resonator. The detector (D) signals were recorded with data acquisition cards (DAC). The quantum cascade lasers (QCL) were driven either by a pulsed or a cw source and the tuning ramp for the laser was generated via the digital-to-analog-converter (D/A).

The QCL was driven and frequency swept using an external trigger signal and voltage ramp (Q-MACS, neoplas control). This is frequently referred to as the *inter* pulse mode (see 4.2.1). Each laser pulse represents a spectral data point in the spectrum. Generally, the spectrum consisted of 160 pulses while the laser was active and 30 points while the laser was off, in order to obtain the offset or zero signal. The repetition frequency of the trigger was set to 10 kHz which enabled the decay transient - being of the order of less than 10 µs - to be followed completely and without any distortion. A fast digitising oscilloscope (LeCroy, Waverunner 104Xi) was used to acquire the ring down transients. All the decays of a single sweep (190 pulses × 100 µs trigger period = 19 ms) were recorded with an appropriate time resolution (sub-µs range) in one shot. In practice the ring down decay was recorded for several µs and the rest of the QCL off phase was not detected until the next trigger event occurred. In this way a data set consisting of 160 single unaveraged decays was obtained. The ring down transients were then processed according to equation (6 - 6) for CRDS and equation (6 - 10) for ICOS, i.e. each ring down transient was analysed twice, firstly with a linear fit to the decay on a semi-logarithmic scale from which τ was obtained, and secondly the area under the transient decay was integrated to obtain the integrated cavity output I. The baseline I_0 for ICOS (required in the data in figures 6.6 and 6.9) was similarly determined by integrating over the output of an empty cavity.

Two pulsed QCLs (Alpes Lasers) were used for the experiments. One emitting between 1345 and 1352 cm^{-1} ($\sim$ 7.42 µm) almost coincided with the centre wavelength of the high reflectivity mirrors, whereas the emission of the second laser between 1195 and 1200 cm^{-1} ($\sim$ 8.35 µm) coincided with the edge of the low loss range of the cavity mirrors. The first QCL was operated at 1.6°C corresponding to an emission sweep from

1347.6 cm^{-1} to 1348.2 cm^{-1}. A pulse width of 60 ns was chosen in order to inject enough power into the cavity. For the 8.35 μm laser a pulse width of 32 ns was adequate to detect a signal from the cavity. For this laser the temperature was set at 4.0°C resulting in a spectral output of 1197.2 to 1197.8 cm^{-1}.

6.3.2   Continuous cavity excitation

The only difference to the optical setup in the previous section and figure 6.1 is that the radiation of the QCL was directed to the telescope with an aspherical ZnSe lens (30 mm diameter, f/1.5). First, the cavity output signal and reference spectra were recorded with the same LN cooled detectors used in the CRDS. Since the sensitivity of the fast detector, formerly sensing the ring down transients, was lower for the current spectral range it was replaced by the detector used formerly as a reference detector. Next, a TE cooled detector (neoplas control, VIGO element PDI-3TE-10/12) was employed for the signal channel. Two different cw QCL (Alpes Lasers) were employed: the system validation was carried out with a QCL (sbcw848) emitting from 1300 cm^{-1} to 1311 cm^{-1}. A very similar replacement laser (sbcw847, 1300 ... 1307 cm^{-1}) was employed for the subsequent sensitivity improvements. The target spectral position had to be slightly shifted for these experiments from ~ 1307 cm^{-1} to ~ 1303 cm^{-1} matching similar strong absorption features as before. Both QCLs were operated in the same standard housing (Starter Kit, Alpes Lasers) and TE cooled.

A DC current source (Kepco, BOP100-2M) provided up to 2 A and 100 V. The first laser (sbcw848) was typically driven with a DC current of around 380 mA. An additional current ramp of 45 mA was impressed on it in order to sweep the laser frequency during the on-phase. The laser current was briefly reduced below the threshold value by -70 mA to record the offset of the zero signal (figure 6.2) in the off-phase. The sweep rate of the laser was 0.9 kHz for all validation measurements.

The spectra (on-phase) consisted of 800 points with an additional off-phase of 50 points. After switching the current from the off to the on-phase the QCL required a short period of about 50 μs to stabilise its temperature and therefore its frequency. After that the laser sweeps 0.78 cm^{-1} (23.4 GHz) with the current modulation applied. This is illustrated at the top of figure 6.2 showing the signal through a Germanium etalon. In combination with stable reference gases (CH$_4$, N$_2$O) this scan also served for the identification and calibration of the frequency axis of the spectra.

For a mirror distance of $L = 0.432$ m, used for all validation experiments, the sweep of the laser frequency corresponds to a scan over 67 cavity modes. In other words, after approximately 14 μs a new cavity mode was excited which was sufficient to avoid an overlap between the decay events of different modes. On the other hand the sweep rate gave a full spectrum after slightly more than 1 ms (figure 6.2). A given number of spectra were then averaged and simultaneously fitted. In order to smooth interference effects on the baseline of the spectra, mainly caused by longitudinal cavity modes, the whole cavity was arbitrarily destabilised mechanically, e.g. by an external vibrating source. Averaging over 5 s or more (i.e. > 1000 single shot spectra) reduced the interference fringes to below the noise level induced by all the data acquisition electronics.

Since the cavity mirrors itself served to enclose the evacuated cavity, which would be also the case for the intended application to plasma diagnostics, a mechanical dither of the mirror mounts by piezo electric transducers could not be implemented. In order to increase the sensitivity, firstly, the cavity length was increased to $L = 1.297$ m to achieve a higher density of excited transverse cavity modes and secondly, the QCL current was electronically dithered. Therefore the replacement laser (sbcw847) was operated with a different high compliance voltage laser controller (ILX Lightwave LDX-3232) which could be externally modulated with a high frequency signal. Similar to the earlier validation experiments the laser current was ramped over 80 mA starting from a DC level of 480 mA to accomplish a frequency sweep of 0.67 cm^{-1} (20.1 GHz). Additionally, a 3 mA (peak-to-peak) sinusoidal modulation at almost 200 kHz (i.e. arbitrarily out of phase with the laser sweep rate of 0.9 kHz) was superimposed. Hence the occurrence of resonance between the laser and (fundamental) cavity modes was electronically dithered. Figure 6.3 illustrates the achieved transmission with an empty cavity for 4 representative sample sweeps. If only a constant DC current is applied to the QCL (figure 6.3 a) the transmitted signal is governed by occasional high intensity resonances which are more pronounced during the first half of the on-phase when the QCL temperature and hence the emission frequency stabilises. As soon as a current ramp is superimposed continuous transmission is observed leading to a smoothed average signal, albeit single resonance events may be still of high intensity (figure 6.36 b). The influence of these spikes is reduced by dithering the QCL current (figure 6.3 c and d). Although a high amplitude of the sinusoidal dither yields apparently a better signal-to-noise ratio (SNR) (figure 6.3 d), the additional modulation had to be limited to 3 mA (peak-to-peak value) (figure 6.3 c). The reason is discussed in section 6.4.3 after the analysis of the pulsed CRDS experiments.

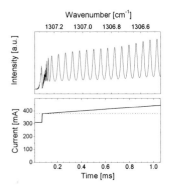

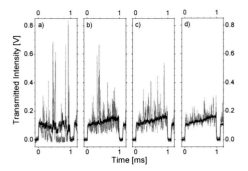

Figure 6.2: **Lower:** Average current (dashed line) and its modulation (heavy black line) applied to the cw QCL (sbcw848) during a single sweep. **Upper:** Signal through a 25.4 mm Germanium etalon (0.048 cm^{-1} FSR) demonstrating the 0.78 cm^{-1} sweep of the laser when the modulation ramp was applied.

Figure 6.3: Improvement of the residual noise level by an electronically dithered QCL (sbcw847) current. The influence is demonstrated for the single shot cavity transmission (grey) and the transmitted signal averaged over 50 laser sweep cycles (black). Detector signals correspond to an empty cavity a) without a ramp or dither (480 mA DC current), b) with an additional current ramp (80 mA), c) with a weakly dithered (3 mA) current ramp, and d) with a strongly dithered (10 mA) current ramp applied.

6.4   Results and Discussion

CRDS and ICOS measurements with two different pulsed QCLs are presented in section 6.4.1. The pulse length was chosen to be as short as possible while maintaining a reasonable cavity output. The achieved sensitivities are compared with conventional multiple pass spectrometers. The parameters which determine the sensitivity are quantitatively discussed and summarised in detail in section 6.4.2. The evaluation of a CEAS system including a discussion on current sensing limits is presented in section 6.4.3.

6.4.1   CRDS and ICOS using a pulsed QCL

6.4.1.1 CH_4 detection at 7.42 μm

The characteristics of the empty cavity were first established (figure 6.4). The average τ_0 across the spectral mode was 900 ns corresponding to a mirror reflectivity of about 99.84 %, which was slightly lower than specified, and an effective absorption path length L_{eff} of 270 m. A relative error, $\Delta \tau / \tau_0$, of about 7.6 % was estimated from the scatter of the τ_0 data. For the case of $L = d$ in (6 - 15) this corresponds to a minimum detectable (peak) absorption coefficient of 2.8×10^{-6} cm^{-1} in a single shot experiment. The theoretical NEA for the ~ 50 Hz laser sweep rate should be 4×10^{-7} cm^{-1}Hz$^{-1/2}$. However, in practice averaging experiments were not undertaken to achieve this theoretical figure.

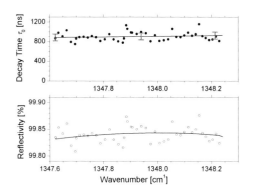

Figure 6.4:
Decay time measured at 7.42 μm for an empty cavity (upper) and the corresponding mirror reflectivity (lower). Solid lines represent a polynomial fit to the data.

When CRDS and ICOS spectra of 100 ppm CH_4 in a background of 300 mbar Ar were recorded deviations from τ_0 or I_0 were observed. For the CRDS spectrum (figure 6.5) the indicated deviations are slightly more pronounced than the scatter of the τ_0 data (figure 6.4), but nevertheless the potential absorption features hardly exceed the noise level. The experimental absorption coefficient in figure 6.5 never exceeds 2×10^{-5} cm^{-1} as shown around 1347.85 - 1347.90 cm^{-1}. The ICOS spectrum (figure 6.6) appears not to be as noisy as for the CRDS case (integration over the cavity output is in fact an averaging process) and the deviation from the baseline I_0 (figure 6.6) is now obvious. The k values from the ICOS spectrum are below 6×10^{-5} cm^{-1}, clearly larger than that obtained from the CRDS analysis but almost a factor of ten less than expected from a calculation using the $HITRAN$ database

(2004 edition) [64]. The two CH_4 lines that should theoretically appear in the spectra under the current conditions are indicated in figures 6.5 and 6.6. Both should be clearly resolved and exhibit an absorption coefficient of nearly 5×10^{-4} cm^{-1}. Hence the shortfall in k of the CRDS results is a factor of 25 or greater. On the other hand the features in figures 6.5 and especially 6.6 suggest broadband absorption, i.e. unresolved absorption of several lines, rather than single CH_4 lines. This potential broad absorption feature appears well below the reference position of one of the single CH_4 lines at 1347.92 cm^{-1} which has been carefully calibrated with the peak of the corresponding CH_4 feature in a reference cell.

Similar observations have been made by Sukhorukov et al. [51]. They concluded that the dual effects of line broadening and line position shifts might be minimised to some extent by reducing the pulse widths used to operate the laser. In our case it is 60 ns and it was found that reducing it lead to an insufficient signal for detection after the cavity.

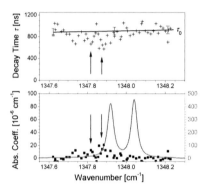

Figure 6.5: CRDS results for 100 ppm CH_4 in Ar at 300 mbar total pressure: decay time τ (upper) and the corresponding absorption coefficient k (lower). Filled squares in the lower panel are the experimental absorption points; solid line is the calculated spectrum. Note the 5 times scale magnification for the calculation (right hand scale). Arrows in the upper and lower panel indicate potential low intensity absorption features.

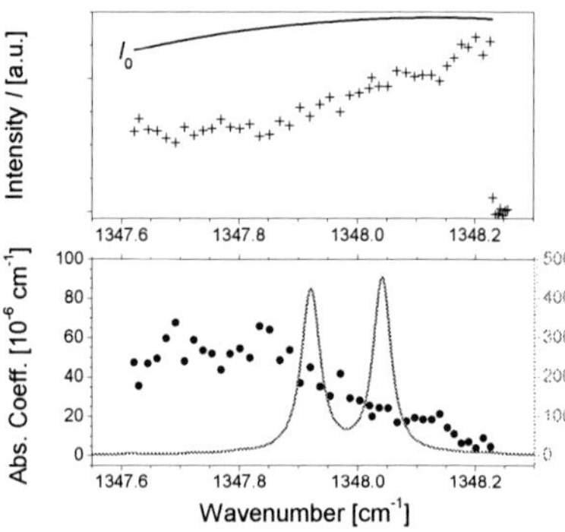

Figure 6.6: ICOS results for 100 ppm CH_4 in Ar at 300 mbar total pressure: the absorption coefficient k (lower) was derived from I_0/I values in the upper spectrum. Filled circles in the lower panel are experimental absorption points, solid line is the calculated spectrum. Note the 5 times scale magnification for the calculation (right hand scale).

6.4.1.2 N_2O detection at 8.35 μm

Since smaller QCL pulse widths might improve the sensitivity and selectivity of the pulsed CRDS experiments it was decided to move to the 8.35 μm region where the mirror reflectivity is lower (transmission > 0.16 %). A pulsed QCL emitting at the edge of the high reflectance regime of the same mirror set was combined with the cavity. This enabled the laser pulse width to be reduced to 32 ns. The decay time of the empty cavity was ~ 500 ns corresponding to an expected drop in the reflectivity to 99.71 % (figure 6.7), i.e. almost quadrupling the intensity transmitted through one mirror. The effective path length L_{eff} falls to 150 m which is now of the same order of magnitude as available from conventional Herriott-type long path cells. The relative error in $\tau \sim 6.5\ \%$ is similar to that reported above for CH_4. For a single shot experiment the minimum detectable (peak) absorption coefficient is $\sim 4.3 \times 10^{-6}\ cm^{-1}$, reduced due to the absorption length. The corresponding NEA in the case of averaging would be $6.1 \times 10^{-7}\ cm^{-1}Hz^{-1/2}$.

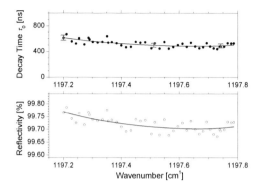

Figure 6.7:
Decay time measured at 8.35 μm for an empty cavity (upper) and the corresponding mirror reflectivity (lower). Solid lines represent polynomial fits to the data.

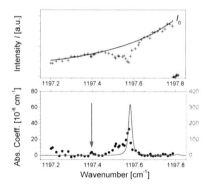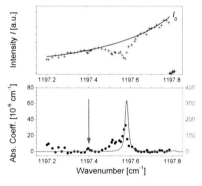

Figure 6.8: CRDS results for 1667 ppm N_2O in N_2 at 100 mbar total pressure: decay time τ (upper) and the corresponding absorption coefficient k (lower). Filled squares in the lower panel are the experimental absorption points; solid line is the calculated spectrum. The arrows indicate two weak N_2O absorptions in the calculated spectrum. Note the 5 times scale magnification for the calculation (right hand scale).

Figure 6.9: ICOS results for 1667 ppm N_2O in N_2 at 100 mbar total pressure: the absorption coefficient k (lower) was derived from I_0/I values in the upper spectrum. Filled circles in the lower panel are experimental absorption points, solid line is the calculated spectrum. The arrow indicates a weak N_2O absorption which is also visible in the experimental spectrum. Note the 5 times scale magnification for the calculation (right hand scale).

Next, measurements were performed with a gas mixture consisting of 1667 ppm N_2O in N_2 at a pressure of 100 mbar. Some improvements in the CRDS spectra (figure 6.8) as well as in the ICOS spectra (figure 6.9) can be detected compared to the previous CH_4 experiments. The spectra exhibit an absorption line but artefacts similar to those described above are still present. An N_2O absorption feature is indicated by an unambiguous drop in the decay. The peak absorption coefficient k derived from CRDS is about 5×10^{-5} cm^{-1} at 1197.5 cm^{-1}. Nevertheless the data points in the spectrum are still fairly scattered. Notably the ICOS spectrum shows a significant improvement compared to all other spectra presented earlier. The absorption coefficient is of the same order of magnitude as deduced from the CRDS experiments. The peak appears at 1197.58 cm^{-1} (figure 6.9) and the broadening mentioned earlier shows up as a shoulder to the left of the absorption line. Although this result is qualitatively closer to the calculated frequency from the *HITRAN* database, it still falls short by at least a factor of 8 in the absorption coefficient.

6.4.1.3   Sensitivity conclusions

It could be argued that both the CH_4 and the N_2O absorption features chosen as examples are far too strong for a standardisation test since saturation effects could arise thereby reducing the experimental absorption coefficients. In this case weaker absorption features would show a better agreement between experiment and calculation. However, considering e.g. the smaller N_2O lines which should appear at 1197.40 cm^{-1} and 1197.81 cm^{-1} (figure 6.8 and 6.9) it is clear that potential saturation effects cannot explain the observed effects. The small N_2O line at 1197.40 cm^{-1} appears above the noise level and agrees with the calculation, but note that the latter is plotted with a times 5 scale magnification. The shortfall in k for weaker or stronger lines is very similar, i.e. the sensitivity is systematically limited.

Provided that spectral averaging was carried out the achieved sensitivity for the pulsed CRDS measurements should be either 4×10^{-7} cm^{-1}Hz$^{-1/2}$ (7.42 μm) or 6×10^{-7} cm^{-1}Hz$^{-1/2}$ (8.35 μm). These values are approximately a factor of 10 better than estimated from similar experiments performed by Manne et al. [50,65] and a factor of 10 worse than ICOS results with a pulsed TE cooled QCL [49,66]. In the latter case a better SNR and a ~ 5 times longer effective path length yielded better sensitivity. For CRDS with a cw QCL a notably higher sensitivity was reported (4×10^{-9} cm^{-1}Hz$^{-1/2}$ [47]) also suggesting a systematic source of error with pulsed QCLs. It is interesting and instructive to compare the *NEA* of the different CRDS measurements with that using a long path cell. For comparison, Menzel et al. used a Herriott cell type QCL spectrometer of path length 100 m. Their *NEA* for short time measurements (~ 1 s) is estimated to be 3×10^{-8} cm^{-1}Hz$^{-1/2}$ [52,67] and would therefore be superior to the high finesse cavity setup used here. More recently Nelson et al. improved the detection limit for a Herriott cell (210 m) combined with a QCL to 3×10^{-10} cm^{-1}Hz$^{-1/2}$ [39,68], i.e. about 3 orders of magnitude better than employing resonant optical cavities with pulsed QCLs and still an order magnitude better than with cw QCLs as the laser source.

To summarise, the main obstacles in the experiments with a pulsed QCL combined with a high finesse optical resonator are: the selectivity is decreased due to arbitrarily broadened and shifted absorption lines and the sensitivity is substantially reduced due to diminished absorption signals in comparison with theory. Clearly, this could be overcome by calibration [50,51], but in so doing one of the major advantages of CRDS, namely the calibration free measurement approach, is lost. Conversely this raises the question as to whether the combination of an optical cavity with a pulsed QCL in order to benefit from reduced sample volumes compared with multiple pass cells of the same effective length is desirable as a calibration free method. The reason for these degradations from the theoretical performance should be accounted for, because the measurements of τ_0 with empty cavities shows potential for obtaining quantitative results.

6.4.2   Bandwidth effects with a pulsed QCL and optical cavities

Possible explanations for the effects observed here can be found in the work of Sukhorukov et al. [51]. They discussed the deviations in their measurement series (i.e. absorption lines with an increased line width and shifted line positions with increasing pulse width), which were not as pronounced as those presented here, in terms of the well-known frequency-down chirp of the QCL (cf. chapter 4). In order to illustrate this, the frequency chirp of the 8.35 μm laser was studied in detail (section 6.4.2.1). The consequences of such bandwidth effects for (pulsed) CRDS are quantitatively discussed and generalised in 6.4.2.2.

Optical feedback due to reflected light from the cavity mirrors does not provide an explanation for the line shift, because the QCL is already tuned out of resonance due to the laser chirp when the reflected light arrives at the laser, about 1 ... 3 ns after the initial emission (for the actual distance between cavity and laser). Multimode behaviour of the lasers might be another explanation for apparent line broadening or line shifts. However, FTIR spectra of the QCL emission showed no evidence for this hypothesis. Furthermore, the simultaneously recorded reference spectra were not affected in the same manner.

6.4.2.1   Determination of the frequency chirp

The frequency chirp of the 8.35 μm laser was evaluated from the direct emission of the QCL (i.e. without the cavity in place and with the current ramp off) modulated with a Germanium etalon of 0.048 cm^{-1} (1.44 GHz) free spectral range (FSR). The signal was recorded with a fast TE cooled detector (IPM Freiburg, Vigo element PDI-2TE-10.6). The pulse length was determined from the recorded laser pulse intensities at the noise or zero signal level. Short pulses of less than 40 ns were mainly affected by oscillations originating from the driving current so that the etalon fringes could not be discriminated. For longer pulses all undistorted fringes were analysed. This resulted in a value for the chirp of the QCL for only a part of the laser pulse, namely that without current oscillations. For simplicity, it was assumed that the chirp rate $d\nu/dt$ of the QCL is constant throughout the laser pulse t_{on} and hence the full chirp of the laser was extrapolated on the basis of the deduced partial chirp value and the measured total pulse length $\Delta\nu(t_{on}) \approx d\nu/dt \cdot t_{on}$. Figure 6.10 shows the full chirp $\Delta\nu(t_{on})$ of the QCL for different combinations of operating voltage and pulse length and is, in fact, a supplement to figure 4.11 (middle panel, trace $\triangle$) where the chirp rate is displayed for only one condition (i.e., close to the threshold, $t_{on} = 150$ ns). In what follows it will be demonstrated that for a combination of pulsed QCLs with optical cavities both the chirp rate and the full chirp are critical parameters.

In order to increase the usable output signal of the cavity in a CRDS experiment the input can be increased either by modifying the driving voltage or the pulse width. Both factors significantly increase the full chirp of the QCL. It can be seen (figure 6.10) that for operating voltages close to threshold with small pulse widths, e.g. 32 ns, as used here, the chirped QCL pulse should nevertheless cover at least 0.1 cm^{-1} (3 GHz). In order to verify this for the same QCL operating conditions as in the previous CRDS measurements a 76.2 mm long Germanium etalon (0.0163 cm^{-1} or 489 MHz FSR) was used. The QCL was swept by a current ramp and each pulse in the train of 160 pulses was recorded. Figure 6.11 shows an example for one pulse in which a full frequency chirp of about 0.16 cm^{-1} (4.8 GHz) was confirmed. The chirp rate of the laser is 0.005 cm^{-1}/ns which accords with the values found for pulsed QCLs in chapter 4 (figures 4.10 and 4.11) and by other groups [50,51].

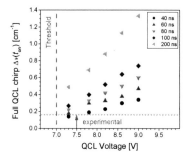

Figure 6.10: Analysis of the full frequency chirp $\Delta\nu(t_{on})$ of the 8.35 µm laser observed for laser pulses of different length t_{on} and intensity, i.e. for different driving voltages. The voltage threshold of the laser is marked with a dashed vertical line. The chirp expected for the operating conditions of the CRDS experiments is indicated by a dotted horizontal line.

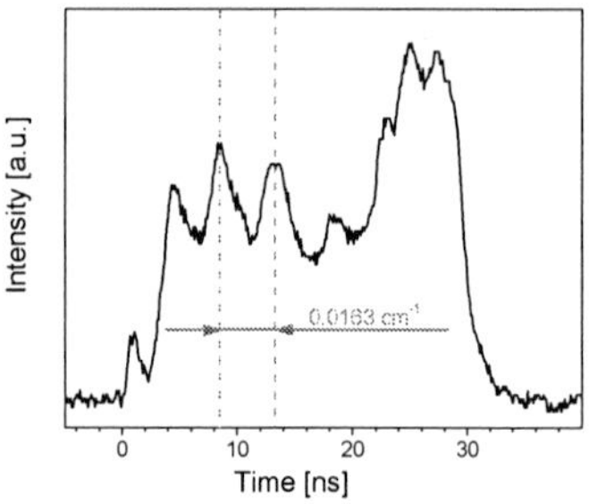

Figure 6.11: Example of the QCL chirp for one 32 ns pulse of the 8.35 µm laser at 90 % of the applied current ramp, displayed with the aid of a 76.2 mm Germanium etalon (0.0163 cm^{-1} or 489 MHz FSR).

6.4.2.2   Consequences of the chirped pulse for optical cavities

The consequences of such a spectrally extended laser pulse are depicted schematically in figure 6.12. A sample absorption feature of 0.03 cm^{-1} (900 MHz) full width at half maximum (FWHM) is shown as might be observed at intermediate or atmospheric pressures. The upper panel in figure 6.12 is an idealised representation in a theoretical high finesse cavity. Two aspects should be noted. Firstly, the mode structure of the 0.432 m long cavity with a free spectral range of 0.012 cm^{-1} (360 MHz) was calculated using the expressions given by Zalicki and Zare [63]. Since the excitation of the cavity with 32 ns (or even the 60 ns pulses) neither matches the special case of being twice the round trip time nor being longer than the decay time the idealised picture of an excitation longer than the decay time has been chosen. This would of course only be true for cw-illumination of the resonator. The arguments about the effects connected with the chirped QCL pulse are, however, not influenced by this choice. Secondly, the laser line width has been set to 0.0012 cm^{-1} (36 MHz) which is typical for a non-chirped cw QCL with modest power supply ripple (cf. 4.3.3). Hence the upper panel of figure 6.12 describes the ideal case for an optical resonator based spectroscopic experiment, where $\Delta \nu_{laser} < \Delta \nu_{FSR,cav} < \Delta \nu_{FWHM,line}$ yielding single exponential decays.

If now the chirped QCL pulse is taken into account the picture has to be modified. Although this situation has been considered earlier by Silva et al. [49] and Sukhorukov et al. [51], the effect is usually underestimated. Silva determined an upper limit for the effective laser line width of their pulsed QCL, from a fit to NO absorption lines, to be 0.024 cm^{-1}. Since the FSR of their cavity and the absorption features were comparable in width to those presented here (figure 6.12) they concluded that the QCL excites several cavity modes but still fewer than the number encompassed by the absorption feature. The ratio between the absorption line width $\Delta \nu_{FWHM,line}$ and the number of excited modes times the FSR of the cavity $N \cdot \Delta \nu_{FSR,cav}$ is the important criterion. Usually the effective laser line width $\Delta \nu_{laser}$ is assumed to give an adequate estimate of the number of affected cavity modes which is not the case.

It was established in chapter 4 (section 4.3) that both the effective line width $\Delta \nu_{laser}$ and the full chirp $\Delta \nu(t_{on})$ strongly depend on the QCL being used and thus on the obtained chirp rate. Using equation (4 - 3) to estimate $\Delta \nu_{laser}$ from the chirp rate of the 8.35 μm QCL used here, would result in an effective laser line width of 0.013 cm^{-1} (390 MHz) for the limiting case of C = 1. In the context of figure 6.12 this would lead to the same conclusion as in Silva's case. However, the expression for $\Delta \nu_{laser}$ was derived for a best aperture time Δt. By adjusting the QCL pulse width to facilitate a detectable cavity output this best aperture time is exceeded. It was also pointed out in section 4.3 that more generally equation (4 - 1) is valid, i.e. the spectral width of a chirped pulse is determined by the pulse width and/or the Fourier transform limited laser line width. Hence, for considerably long QCL pulses, the chirp set by the pulse width is the dominating factor. For the 8.35 μm QCL the full chirp $\Delta \nu(t_{on})$ covers 0.16 cm^{-1} (4.8 GHz) which drastically changes the picture as shown in the bottom panel of figure 6.12. With a single laser pulse almost twice the number of cavity modes are excited than are affected by a representative broad absorption feature at intermediate pressure ≥ 50 mbar.

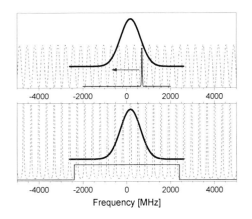

-4000 -2000 0 2000 4000

-4000 -2000 0 2000 4000
Frequency [MHz]

Figure 6.12:
Simulation of the spectrum of a 0.432 m long cavity (FSR 360 MHz, dashed) combined with a molecular absorption line (FWHM 900 MHz, solid thick) and the idealised frequency coverage of the QCL emission (solid narrow line) with a laser line width of 36 MHz (upper), and for the experimentally observed case with a frequency chirp covering 4.8 GHz (lower).

The consequences are i) non-exponential decays, ii) reduced absorptive losses, since a part of the injected power is transmitted via undamped cavity modes, and iii) an apparent line broadening since the absorption feature is present in the decay signal for a much larger number of pulses. By analogy to section 4.4 each pulse or decay signal corresponds to a measurement channel and determining the decay constant over μs is an averaging process that reduces the available analogue detection bandwidth. This may finally also lead to artefacts such as a shoulder close to the main absorption feature.

All these features are apparent in the results described in section 6.4.1. Laser bandwidth effects of this type have been described extensively by Hodges et al. [69] for a 3 GHz line width dye laser used to probe a 900 MHz (FWHM) absorption feature. They observed 5 to 8 times smaller absorptive losses compared to using a Ti:Al₂O₃ laser source having a bandwidth smaller than the absorption line. This reduction in sensitivity accords well with the approximately 10 times smaller absorption coefficients reported in section 6.4.1.

Berden et al. discussed laser bandwidth effects, the reason for non-exponential decays as well as underestimated absorption coefficients in a review [70]. They also pointed out that correct values might be extracted if the spectral density distribution of the light source is known. The investigation of individual pulses during a laser sweep in the *inter* pulse mode in chapter 4 (sub-section 4.4.4), however, leads to the conclusion that the determination of the spectral density distribution of the QCL pulse is not an option due to the unconstant pulse width and laser intensity throughout the sweep. This would additionally be complicated by the variable chirp rate during an individual QCL pulse (sub-section 4.3.1).

The conclusion from this section is that an infrared source with much smaller effective line width or spectral coverage is needed to reach the ideal case of a high finesse cavity experiment suggested by the upper panel in figure 6.12. Since the frequency-down chirp is inherent to pulsed QCLs the only solution appears to be the use of cw QCLs. It should be pointed out that the chirped laser pulse is counter productive in two ways that may be synergistic. Firstly, the chirp rate of 150 MHz/ns causes an insufficient cavity build-up because each cavity mode only coincides with the corresponding spectral part of the laser emission for a very short time. This corresponds in fact to a laser pulse at a fixed frequency shorter than the cavity round trip time propagating in the cavity. The throughput is then limited by the reflectivity of the mirrors [63]. For slightly longer interaction with the cavity the intensity gain in the developing cavity modes is still rather limited and so is the output power of the cavity. Secondly, the full chirp of the QCL excites too many resonator modes as discussed above. In order to achieve a higher cavity output the pulse length or the operating voltage has to be raised which in turn increases the chirp rate and the full chirp respectively, so even more cavity modes are excited during one laser pulse while the interaction time may even be reduced.

6.4.3   CEAS using a cw QCL

6.4.3.1   Calibration of the method and test of validity

A room temperature cw QCL combined with the optical resonator used earlier was applied to perform CEAS experiments. Due to small instabilities in the current source for the QCL the (model) laser line width (FWHM) of 0.0012 cm⁻¹ (36 MHz) suggested in the top panel of figure 6.12 is expected to be a lower limit in practice[2]. From fits to the spectra recorded in experiments without electronically dithering the QCL current an upper limit (FWHM) of the instrumental broadening of 0.010 cm^{-1} (300 MHz) was deduced. This reflects mainly the longitudinal mode spacing of the short cavity ($\Delta \nu_{FSR,cav}$ = 0.012 cm^{-1} or 360 MHz) rather than effects of the data acquisition approach etc. However, mechanical destabilisation already caused the excitation of additional transverse cavity modes. Hence the laser (being swept by means of a current ramp) probes the absorption feature with a spectral resolution slightly better than $\Delta \nu_{FSR,cav}$ which is sufficient to fall below the absorption line width.

Spectra of N_2O from different mixtures were recorded, in order to determine the effective reflectivity of the cavity mirrors, and provide a calibration. The measurements were taken under flowing conditions at a constant pressure of 1.2 mbar while the N_2 buffer gas and N_2O flow was varied for the different N_2O calibrations. Figure 6.13 shows the result of the spectral scans, obtained with a LN cooled detector and averaged over 20 s. Lower intensity N_2O lines, just exceeding the noise level ~ 2 × 10^{-3} in transmission, can be seen between the three dominant N_2O features in the spectrum at 100 ppm. The line positions and line strengths from the *HITRAN* database are given in table 6.1 [64]. The relative absorption of the two lines with a line strength of nearly 10^{-20} cm/molecule was 17 % in the 100 ppm N_2O sample. Even for such comparatively small absorptions a correction for non-linearity is recommended (see below). A fit of the transmission spectra resulted in an effective reflectivity of R = 99.96 % for the cavity mirrors or an effective path length of 1080 m. Since in this approach the excitation of high order transverse modes may occur which could have different diffraction losses this value of R should be regarded as an effective one [52].

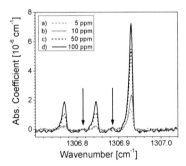

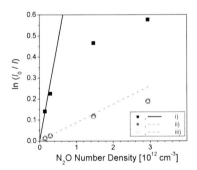

Figure 6.13: CEAS scans for different N_2O calibration mixtures (N_2O diluted in N_2) at a constant total pressure of 1.2 mbar: a) 5 ppm, b) 10 ppm, c) 50 ppm, d) 100 ppm The acquisition time was 20 s. The QCL was swept from high to low wavenumbers. The weak absorption features indicated by arrows appear above noise level only in the spectrum at 100 ppm.

Figure 6.14: Calibration data for the maxima of the three absorption lines and the four mixing ratios of N_2O mentioned in figure 6.13:
i) 1306.929 cm^{-1}, ii) 1306.846 cm^{-1} and iii) 1306.771 cm^{-1}.

---

[2] This assumption is valid for both power supplies (Kepco BOP and Lightwave ILX), although the discussion here firstly concerns the Kepco supply.

Line Position [cm^{-1}]	Line Strength S [cm/molecule]
1306.7712 *	9.72×10^{-21}
1306.8157	6.13×10^{-22}
1306.8461 *	9.72×10^{-21}
1306.8871	5.36×10^{-22}
1306.8874	5.36×10^{-22}
1306.9291 *	1.04×10^{-19}

Table 6.1:
N_2O line strengths and line positions observed in the calibration experiments. The data indicated with (*) were used for further analysis.

The three strong N_2O lines visible in figure 6.13 have been analysed further using equation (6 - 11) which provides a means of checking the validity of the weak absorber assumption. For this purpose $\ln(I_0/I) \approx GU$ at the maximum was plotted against the concentration of N_2O in the cavity (figure 6.14). Due to its order of magnitude higher line strength compared to the other lines, the data for the line at 1306.93 cm^{-1} is non-linear even for number densities in the 10^{11} cm^{-3} range. For the two smaller lines, which have approximately the same line strength, the linear assumption is valid at least up to $\sim 1 \times 10^{12}$ cm^{-3}. Generally, it can be concluded that the linear approximation is valid for $GU \leq 0.15$ whilst for absorption features bigger than 15 % a correction is necessary.

As pointed out earlier, no evidence was found that optical feedback from the cavity to the QCL caused line shifts or perturbed lines. Nevertheless a weak asymmetry can be detected in the absorption profiles displayed in figure 6.13. A potential explanation is the relatively rapid laser sweep used here yielding a full sweep of almost 0.8 cm^{-1} (24 GHz) in 1.1 ms (figure 6.2). By choosing this sweep rate a reasonable SNR for a 1 s integration time was accomplished. Simultaneously the excitation of a neighbouring longitudinal cavity mode was estimated to occur after about five times the ring down time τ_0 which was assumed to be reasonably long for obviating overlapping ring down events. However, such rapid sweep rates result in relatively short ring-up times (being the time when the QCL frequency coincides with a cavity mode which was presently about 390 ns). Firstly, the cavity throughput becomes limited. This has been modelled and discussed by Remy et al. for the analogous case of a fixed laser frequency and a moving cavity mirror [35]. Secondly, a ring-up time being short compared to the ring down event obviously leads to asymmetric line shapes. Any change of the absorption cross section inside the cavity, e.g. by pronounced absorption lines (marked with * in table 6.1), is observed as an unambiguous intensity change within a very few spectral data points when the laser frequency approaches the absorption feature and as an additionally broadened wing when the laser frequency passed the line maximum (figure 6.13). Note that the spectral scan was carried out from high to low wavenumbers here. Since the CEAS experiments require a calibration in any case, such effects on quantitative results are included in the effective reflectivity R or in L_{eff}, but the sweep rate dependence should always be checked carefully. Consequently, a compromise has to be found between achieving a high SNR using short integration times and reducing non-linear effects on the line shapes.

The application of optical resonators in absorption spectroscopy under low pressure conditions might also suffer from non-linear absorption, e.g., power saturation (chapter 2, sub-section 2.2.5) and has been observed for pulsed and cw CRDS respectively [71,72].

Provided that a fraction of 10^{-3} of the QCL intensity of ~ 10 mW is stored in one of the (longitudinal) cavity modes after the laser was in resonance for approximately 390 ns for the present conditions, the Rabi frequency Ω_{lu} (eq. (2 - 28)) would be about 3 kHz for the strongest line in table 6.1 [73]. Considering the idealised case that the transition is pumped twice per round trip, i.e. about 2500 times during the ring down time, and assuming further that this would increase Ω_{lu} in the same order of magnitude, being now 7.5 MHz, the saturation parameter Σ (eq. (2 - 27)) would be almost 2 for the selected transition at 2 mbar [73]. The absorption coefficient for an inhomogeneously broadened line would then be underestimated by ~ 40 % (eq. (2 -32)). In other words, the experimental conditions might be already sufficient to generate non-linear absorption effects, although the laser is swept rapidly and the intensity build-up inside the cavity is definitely incomplete compared to cw CRDS experiments. The next validation experiment (without correction) suggests that the calculation above represents an upper limit and power saturation is still negligible in the present case. However, further investigations, with e.g. defined changes of the QCL power, are necessary in the future.

Finally the CEAS system was validated with measurements on constituents of laboratory air, namely CH_4, N_2O and H_2O, which could be detected within one spectral scan. The spectrum expected to arise for the average mixing ratios of the three molecules was calculated and is plotted in the upper panel of figure 6.15. Some non-interfering absorption features were selected for analysis and these are summarised in table 6.2. Since the water lines are of low line strengths the large amounts of H_2O in air compared to the other trace gases could be examined. The single water line in the spectrum consists of two unresolved lines with a spacing of 0.004 cm^{-1}, so that an effective value for S (for simplicity the sum) is used. A third line shifted by ~ 0.010 cm^{-1} with a line strength of only 20 % of the effective S of the unresolved neighbour appears as a shoulder to the left of the main line(s). This transition was deconvoluted from the main lines and was not used for the quantitative analysis. The transmission spectrum recorded with a TE cooled detector is shown in the lower panel of figure 6.15. The measurements were performed under flowing and with low pressure conditions at 2.2 mbar in order to increase line selectivity. The spectrum was averaged for 20 s. Qualitatively it agrees very well with the simulation.

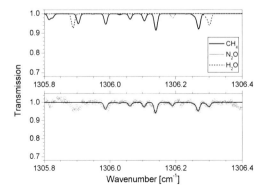

Figure 6.15:
Calculated spectrum (upper) for air containing CH_4, N_2O and H_2O. Experimental conditions: $p = 2.2$ mbar, $L_{eff} = 1080$ m, 0.010 cm^{-1} instrumental broadening (FWHM). The corresponding CEAS spectrum (lower) was observed with a TE cooled detector (open circles) and fitted (solid line) to determine the actual concentrations.

The line positions were fitted to a transmission profile which is also plotted in figure 6.15. This fit was then used to calculate the absorption coefficient using (6 - 10). Integrating over the absorption line of interest revealed the number density and mixing ratio. The results of this quantitative analysis are given in table 6.2: 1.7 ppm of CH_4, 350 ppb of N_2O and more than 1 % of water. A comparison with literature values for the abundance of these molecules in air (1.75 ppm for CH_4, 320 ppb for N_2O and 0.1 - 4 % for H_2O) yields fair agreement [1,74].

6.4.3.2   System performance

In order to be able to evaluate the sensitivity of the TE cooled system the measurements on laboratory air were repeated with a LN cooled detector and the results were very similar. Furthermore, the relative measurement error after 20 s was almost the same ($\Delta I/I_0$ of 0.3 % for the TE detector, 0.2 % for the LN cooled detector). The corresponding *NEA* is then 1.8×10^{-7} cm^{-1}Hz$^{-1/2}$ for the TE cooled detector if the calibration error in *R* is included as given by equation (6 - 17). The system was aligned on-axis with respect to the laser beam which produces cavity mode noise inherent to ICOS/CEAS setups [57]. In order to achieve an accurate SNR, a measurement interval of at least 5 s is necessary to smooth out the residual mode structure on the baseline.

It is possible to estimate the minimum detectable number density (MDND) n_{min} for each molecule by means of equation (2 - 16) assuming that $\ln(I_0/I)$ at the centre frequency ν_0 corresponds to the maximum of the smallest absorption feature that can be distinguished from the noise. The MDNDs of the different molecules analysed are listed in table 6.2. For N_2O a detection limit of 6.4×10^8 cm^{-3} is obtained, which transferred to CH_4 corresponds to 1.6×10^9 cm^{-3} for the strongest line in CH_4. If this is converted into a mixing ratio at 2.2 mbar, where the TE spectrum was acquired, the limits of detection for N_2O and CH_4 are 12 ppb and 46 ppb respectively for a 20 s measurement interval. Using the LN cooled detector these limits would be reduced by a factor of two thirds due to a reduced $\Delta I/I_0$. Clearly for measurements at intermediate or atmospheric pressures sub-ppb levels can be measured although the lines would be broader and might be unresolved.

Molecule	ν [cm^{-1}]	Line Strength S [cm/molecule]	Mixing Ratio	Numb. Dens. [cm^{-3}]	MDND (n_{min}) [cm^{-3}]
CH_4	1305.987	4.78×10^{-20}	1.72 ppm	9.2×10^{10}	2.5×10^9
CH_4	1306.062	2.41×10^{-20}	1.82 ppm	9.8×10^{10}	4.9×10^9
CH_4	1306.105	3.19×10^{-20}	1.68 ppm	9.0×10^{10}	3.7×10^9
CH_4	1306.140	7.60×10^{-20}	1.66 ppm	8.9×10^{10}	1.6×10^9
N_2O	1306.191	1.12×10^{-19}	346 ppb	1.9×10^{10}	6.4×10^8
H_2O	1306.29	6.08×10^{-24} [a]	1.1 %	5.8×10^{14}	1.8×10^{13}

Table 6.2:   Species detected in ambient laboratory air at 2.2 mbar.

[a] effective line strength from two unresolved lines

6.4.3.3   Sensitivity improvements

Cavity mode noise was established as the main limiting factor of the CEAS system used in the previous sub-sections. Although suddenly excited longitudinal modes capable of transmitting high intensities through the resonator are inherent to an on-axis alignment, their influence on the SNR can efficiently be reduced by averaging over theses events. For this reason i) the QCL current was electronically dithered, and ii) the cavity base length was increased.

Dithering the current through the laser enables a better control of the randomised cavity excitation events than the mechanical destabilisation of the resonator used in the earlier validation measurements. It was found in preliminary experiments using an empty cavity (figure 6.3) that a relatively high dither amplitude (> 3 mA or > 0.5 % of the DC current) can significantly improve the SNR. However, employing this mode of operation introduces new artefacts as soon as absorption features appear in a spectrum, e.g., absorption lines exhibit shoulders, are broadened and fractional absorption is reduced. These obstacles are not only similar to those described for the pulsed CRDS experiments (sub-section 6.4.2) - they are of the same origin. The QCL emission frequency is now deliberately chirped which reduces the probability for a resonance of QCL emission and longitudinal modes and thus decreases the mode noise level. On the other hand too many cavity modes are excited simultaneously.

The amplitude of the sinusoidal dither was therefore limited to 3 mA for the subsequent measurements with the replacement QCL (sbcw847) resulting in an uncorrelated dither of slightly more than one cavity FSR per spectral data point. Note that in contrast to the previous experiments the FSR of the cavity was reduced to 0.004 cm^{-1} (120 MHz) by increasing the base length to $d = L = 1.297$ m. Figure 6.16 shows a sample spectrum recorded with 1 s integration time at the new spectral micro-window (1303.1 ... 1303.8 cm^{-1}). An instrumental broadening (FWHM) of 0.008 cm^{-1} (240 MHz) was deduced from a fit to the N_2O feature at 1303.201 cm^{-1}. While the instrumental broadening of earlier measurements using a short cavity mainly reflected the cavity FSR (cf. 6.4.3.1), it is now also affected by the QCL current dither and the rapid laser sweep rate. Since the spectral position was only shifted by 3 cm^{-1}, which is negligible in respect to the entire high reflectance regime of the cavity mirrors (~ 200 cm^{-1}), the same, already established (effective) reflectivity value could be expected.

This was confirmed by a calibration carried out by means of 1.23 ppm N_2O in a standardised gas mixture at different pressures and under flowing gas conditions (figure 6.17). Using the N_2O line indicated in figure 6.16[3] ($S = 1.419 \times 10^{-19}$ cm/molecule) and working now only in the weak absorption regime a reflectivity of 99.967 % was extracted from the slope of $\ln(I_0/I)$ as a function of $n(N_2O)$ (figure 6.17). This yields an effective absorption path of $L_{eff} = 4$ km.

For further sensitivity analysis the Allan variance [75,76] was derived from spectra (similar to figure 6.16) acquired with 1 s integration times over about 30 minutes. The Allan variance was calculated for three selected spectral positions which are indicated in figure 6.16, namely the single N_2O line (●) and two baseline positions (▷, ◁) where no potential absorption features are present. The results are displayed in figures 6.18 and 6.19 respectively.

---

[3]   The two weaker N_2O lines that are also indicated result from unresolved lines and were comparable to the noise level. For this reason they were excluded from further analysis.

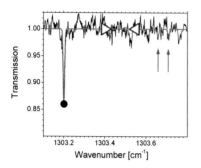

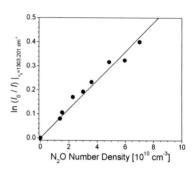

Figure 6.16:   CEAS transmission spectrum of a N$_2$O calibration mixture (1.23 ppm in N$_2$, 0.47 mbar) recorded with 1 s integration time. The N$_2$O line (●, 1303.201 cm^{-1}) and both highlighted spectral positions (▷, 1303.40 cm^{-1} and ◁, 1303.55 cm^{-1}) were used for further analysis. Weak, unresolved N$_2$O features are indicated by arrows.

Figure 6.17:   Calibration data for the N$_2$O line (●, 1303.201 cm^{-1}), mentioned in figure 6.16, acquired in a pressure range from 0.47 to 2.33 mbar. The slope of the linear fit corresponds to an (effective) mirror reflectivity of 99.967 %.

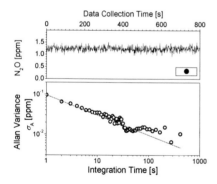

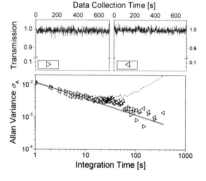

Figure 6.18:   **Upper:** N$_2$O mixing ratio determined at 1303.201 cm^{-1} (● in fig. 6.16) under flowing gas conditions at 0.47 mbar. **Lower:** Allan variance (σ_A) of the upper panel (open circles) and the corresponding white noise trace (grey line).

Figure 6.19:   Baseline signal detected at 1303.40 cm^{-1} (upper left, ▷ in fig. 6.16) and 1303.55 cm^{-1} (upper right, ◁ in fig. 6.16). **Lower:** Allan variance σ_A of the baseline noise for both upper panels (symbols) and the corresponding white noise trace (grey line). An example for an unstable experiment (data stream not shown) is also given (dotted).

An N_2O detection limit (σ_A) of 96 ppb, 17 ppb and 12 ppb at 0.47 mbar is detected in figure 6.18 for integration times of 1 s, 30 s and 90 s respectively. The 1 s value is already interesting for time resolved measurements (plasma diagnostics, online breath gas analysis) whereas the long-time integration limits may be applicable to trace gas measurements. Averaging longer than 30 s is often limited in practice due to drift effects inherent to the system (see below). Hence, the 90 s detection limit should be considered as a best case value of the current system.

In order to facilitate general sensitivity conclusions for the CEAS system the (baseline) noise signal was analysed for the two spectral positions 1303.40 cm^{-1} and 1303.55 cm^{-1} by means of an Allan plot (figure 6.19). The Allan variance σ_A shows the same behaviour in both cases and is mainly determined by white noise. Detection limits of 1.1×10^{-2}, 2×10^{-3}, and 1×10^{-3} are found for the three selected integration times used above. After 30 s integration time small deviations of experimental data from the theoretical white noise behaviour are observed which might be caused by drift effects. The reason for the drift can be directly (i.e. mechanically) and indirectly in nature. Particularly the latter case may hamper further averaging: a thermally induced drift of the laser intensity causes a gradual change of the acquired background normalised transmission spectra and is thus visible as an strong increase of σ_A for integration times longer than 60 s (dotted line in figure 6.19). Additionally, the spectral position will shift and would require a correction.

The Allan variance σ_A of the baseline signal(s) can readily be converted into the absorption coefficient uncertainty Δk by means of equation (6 - 16) and subsequently into the *NEA* using (6 - 17) which would be $\sim 4 \times 10^{-8}$ cm^{-1}Hz$^{-1/2}$ here. Furthermore, if $\sigma_A \approx \Delta I/I_0$ is considered as the minimum detectable peak absorption at the centre frequency ν_0 the MDND (n_{min}) can be inferred (for low pressure conditions) using equation (2 - 16). Note that the instrumental broadening $\Delta\nu_{obs}$ (HWHM) should be used instead of the theoretical Doppler broadening $\Delta\nu_D$ in (2 - 16). In table 6.3 the (anticipated) MDND is collected for several molecules of main interest in plasmas or atmospheric chemistry for three integration times. The given line strengths are typical values for absorption features in the present spectral range [64].

| Species | Line Strength S | Detection Limits (MDND) n_{min} [cm^{-3}] | | |
| | | | Integration Time | |
	[cm/molecule]	1 s	30 s	90 s
N_2O	1.5×10^{-19}	1.6×10^{9}	3×10^{8}	2×10^{8}
CH_4 [a]	5.0×10^{-20}	4.7×10^{9}	9×10^{8}	5×10^{8}
C_2H_2 [a]	1.0×10^{-19}	2.3×10^{9}	4×10^{8}	2×10^{8}
HNO_3 [a]	1.0×10^{-20}	2.3×10^{10}	4.5×10^{9}	2.6×10^{9}
H_2O_2 [a]	2.0×10^{-20}	1.2×10^{10}	2.2×10^{9}	1.3×10^{9}

Table 6.3: Detection limits for molecules being detectable at ~ 1303 cm^{-1} (7.67 µm) with the present CEAS system. An instrumental broadening of 0.008 cm^{-1} (240 MHz, FWHM) and an absorption path of 4 km was assumed for calculating the anticipated values.

[a] anticipated values

6.4.3.4   Discussion of sensitivity achievements

The sensitivity achieved with the CEAS system is comparable to other QCL based high finesse cavity spectrometers reported in the literature. The main parameters of those and the CEAS system studied here are collected in table D.1 (appendix D). To facilitate intercomparisons the specifications of the different systems were converted into the NEA and the MDND for the different molecules. The MDND typically represents the best reported value, sometimes achieved with extended averaging, e.g. determined from Allan variance plots. In some cases the scaling of the NEA with $f^{-1/2}$ according to (6 - 17) does not hold. Therefore the given NEA values should be considered as estimates for short term measurements (~ 1 s) and may not be directly converted in all cases to the MDND.

The sensitivities of 1.8×10^{-7} cm^{-1}Hz$^{-1/2}$ presently achieved with the entirely TE cooled short cavity and of 4×10^{-8} cm^{-1}Hz$^{-1/2}$ for the long cavity CEAS system, respectively, demonstrate a better performance than CRDS setups with pulsed QCLs as discussed in section 6.4.1.3. They are comparable to several other CEAS or ICOS studies exhibiting a NEA down to 4×10^{-8} cm^{-1}Hz$^{-1/2}$ [52,54,56,57,60]. The long cavity setup is still a factor of ~ 10 less sensitive than CRDS employing a cw QCL as demonstrated by Paldus et al. [47] which may be explained by the commonly omitted cavity mode noise in CRDS. For CEAS or ICOS the minimum detectable absorption is typically limited to 10^{-2} to 10^{-3} due to incomplete averaging over the cavity resonances. In combination with wavelength modulation techniques this can be improved by a factor of about 5 [54,56] at the expense of a more complex experimental layout.

Another approach to increase the SNR in the spectra is an off-axis alignment requiring mirrors of bigger diameter. A sensitivity of 2×10^{-9} cm^{-1}Hz$^{-1/2}$ was reported with a cavity base length L less than twice as long as presented here [53]. The MDND achieved for CH$_4$ at 7.9 µm with this method was 2.9×10^9 cm^{-3} compared to 1.6×10^9 cm^{-3} ($< 10^9$ cm^{-3}) for the short (long) resonator in the present study, due to an advantageous spectral position and a lower sampling pressure in our experiment. Recently, an off-axis ICOS spectrometer achieving 9×10^{-10} cm^{-1}Hz$^{-1/2}$ (per scan) has been reported [61].

Not surprisingly the sensitivity of the system described here cannot compete with the values obtained with sophisticated locked CEAS setups (8×10^{-10} cm^{-1}Hz$^{-1/2}$) or even with NICE-OHMS (9.7×10^{-11} cm^{-1}Hz$^{-1/2}$) [48]. Apart from the limitations in the SNR inherent to the CEAS approach, the currently achieved detection limits could be further improved using the nearly 10 times stronger absorption band at 4.5 µm for N$_2$O or the 3 µm region for CH$_4$. The latter spectral range requires interband cascade lasers that have recently become available.

QCL spectrometers employing conventional long path cells which are today increasingly based on TE cooled lasers and detectors accomplish sensitivities well below 1×10^{-7} cm^{-1}Hz$^{-1/2}$ (table D.1) even down to 3×10^{-10} cm^{-1}Hz$^{-1/2}$ [39,68] and hence are still superior to short base length CEAS or ICOS setups. The main reason is the residual mode noise which is absent in the spectra for those spectrometers. However, one aim has been to achieve maintenance free operation with a small sample volume, presently around 0.3 l for the short cavity[4], and with an optical geometry reduced in its complexity which would make the system field-deployable. The volume of the multiple pass cells listed in table D.1, covering effective path lengths up to 210 m, typically ranges from 0.5 to 5 l which may increase the dimensions of the system and the necessary pumping time. More recently, photoacoustic spectroscopy (PAS) has been reported using LN and TE cooled cw QCL [42] or external cavity lasers [41]. Detection limits in the order of $\sim 10^{-7}$ cm^{-1}Hz$^{-1/2}$, but still inferior to methods applying optical cavities, were obtained (table D.1). Moreover, the acquisition time for photoacoustic spectra is rather long ($\sim$ minutes), which does not enable real-time measurements to be performed.

6.5   Conclusions

Thermoelectrically cooled QCLs have been combined with high finesse optical resonators in order to profit from their enlarged path lengths at reasonably small sample volumes in combination with the high absorption cross section in the infrared molecular fingerprint region. Two different approaches have been investigated. Firstly, pulsed QCLs at 7.42 μm and 8.35 μm were used to perform CRDS and ICOS experiments. The spectra measured at pressures between 100 mbar and 300 mbar were normally characterised by a broadening of the absorption lines and reduced absorption in comparison with theoretical expectations, which makes an absolute calibration necessary. The resulting decrease in sensitivity, i.e. to $\sim 5 \times 10^{-7}$ cm^{-1}Hz$^{-1/2}$, means that long path cell configurations of similar sample volumes have superior sensitivity to either CRDS or ICOS. It transpires that the frequency-down chirp inherent to pulsed QCLs sets the fundamental limit. Due to the chirp the effective line width of the QCL is much broader than the narrow molecular absorption features and the rather fast chirp rate does not allow for an efficient build-up of the laser field in the cavity. CRDS using pulsed QCLs therefore has only a limited number of useful applications, e.g. for the determination of the reflectivity of the cavity mirrors in preliminary experiments or for the detection of complex and broad molecular absorptions at higher pressures. Secondly, cw QCLs at 7.66 μm have been combined with an unstabilised and unlocked cavity. With this straightforward arrangement, comprising only a TE cooled detector, an effective path length of 1080 m has been achieved. By increasing the cavity base length an effective absorption path of 4 km and a reduced instrumental broadening due to a reduced cavity FSR were obtained. The main limit to sensitivity in both configurations, of 2×10^{-7} cm^{-1}Hz$^{-1/2}$ or 4×10^{-8} cm^{-1}Hz$^{-1/2}$, was the cavity mode noise which may be reduced by an off-axis alignment. Electronically dithering the QCL current efficiently suppresses a portion of the residual mode noise but increases the instrumental broadening of the CEAS system. Such mode noise is absent in conventional long path cell spectrometers which hence exhibit a higher sensitivity.

   With a 20 s measurement interval detection limits for N_2O and CH_4 of 6×10^{8} cm^{-3} and 2×10^{9} cm^{-3} respectively could be achieved at 2.2 mbar indicating that sub-ppb levels could easily be measured by means of the short resonator at higher or elevated pressures. Furthermore, the small sample volume of 0.3 l used here, and consequently the reduced

---

[4]   The long cavity covered a volume of not much less than 1.5 l.

pumping requirement, is an advantage over long path cells with much larger volumes than 0.5 l.

The detection limits achieved here are relevant both for the detection of radicals in plasma chemistry and for trace gas measurements with field-deployable systems. Additionally, trace gas measurements may also be carried out without pre-concentration procedures. Radicals with small abundances in the gas phase might now be detectable via QCL based spectroscopy. The choice of an appropriate method depends on whether the important criterion is ultimate sensitivity or a more compact system. In the former case a multi pass cell spectrometer would be preferable because of its better signal to noise characteristics but this configuration would exclude the detection of processes on short time scales or certain types of *in-situ* measurements. Sophisticated locked cavity schemes might overcome such restrictions at the expense of complex spectrometer geometries. To achieve a compact system a small volume cavity based spectrometer employing cw QCLs would be more appropriate. This configuration would also be of special interest for applications where the pressure cannot arbitrarily be chosen in order to adapt the absorption line width to the laser line width or instrumental broadening, e.g. in low pressure plasmas. Moreover, for these applications *in-situ* measurements are essential because multi-pass cell sampling *ex-situ* is not an option.

D Appendix

Ref.	Method	λ [μm]	QCL	Operating Temperature	Detector Cooling	L_{eff} [m]	NEA [a] $[\text{cm}^{-1}\text{Hz}^{-1/2}]$	Gas	$MDND$ [b] $[\text{cm}^{-3}]$	Recording Time [s]
[50,65]	CRDS	10.3	pulsed	near ambient	cryogenic	240	5×10^{-6}	NH_3	1.2×10^{12}	20
[55,77]	CRDS	5.2	cw	cryogenic	cryogenic	[a] 1050	1×10^{-7}	NO	1.4×10^{9}	8
[47,78]	CRDS	8.5	cw	cryogenic	cryogenic	[a] 280	[c] 4×10^{-9}	NH_3	6.7×10^{9}	1
[52,79]	CEAS	5.2	cw	cryogenic	TE	670	3×10^{-6}	NO	1.6×10^{10}	200
[60,80]	ICOS (off-axis)	3.5	cw[d]	cryogenic	cryogenic	[a] 83	5×10^{-7}	H_2CO	1.6×10^{11}	3
[54,81]	ICOS[e]	5.2	cw	cryogenic	cryogenic	75	2×10^{-7}	NO	3.3×10^{10}	15
This work	*CEAS*	*7.7*	*cw*	*near ambient*	*TE*	*1080*	*2×10^{-7}*	*CH_4*	*1.6×10^{9}*	*20*
This work				*near ambient*				*N_2O*	*6×10^{8}*	*20*
This work	*ICOS[e]*	*5.5*		*near ambient*	*cryogenic*	*4000*	*4×10^{-8}*	*N_2O*	*2×10^{8}*	*90*
[56,82]	ICOS[e]	5.5	cw	near ambient	cryogenic	700	1×10^{-7}	NO	2.1×10^{10}	1
[49,66]	ICOS	5.2	pulsed	near ambient	cryogenic	1500	6×10^{-8}	NO	$\sim2\times10^{9}$	4
[57,83]	ICOS	5.5	cw	near ambient	cryogenic	500	4×10^{-8}	NO	1.2×10^{10}	4
[53,84]	ICOS (off-axis)	7.9	cw	cryogenic	cryogenic	[a] 5560	[c] 2×10^{-9}	CH_4	2.9×10^{9}	10
[61]	ICOS (off-axis)	6.7	cw	cryogenic	cryogenic	4200	[c] 9×10^{-10}	HDO	2.6×10^{8}	4
[48]	CEAS (locked)	8.7	cw	cryogenic (?)			[c] 8×10^{-10}	N_2O		
[48]	NICE-OHMS	8.7	cw	cryogenic (?)			[c] 1×10^{-10}	N_2O		
[45,85]	QCLAS	4.3	pulsed	near ambient	TE	56	1×10^{-7}	CO_2	5.3×10^{8}	240
[52,67]	QCLAS	5.2	cw	cryogenic	TE	100	3×10^{-8}	NO	4.2×10^{9}	200
[44,86]	QCLAS	7.9	quasi-cw	cryogenic	cryogenic	100	2×10^{-8}	CH_4	$\sim2\times10^{9}$	30
								N_2O	$\sim7\times10^{8}$	30
[37,87]	QCLAS	7.9	pulsed	near ambient	TE	56	2×10^{-8}	CH_4	1.1×10^{9}	200
								N_2O	5.9×10^{9}	1
[37,88]	QCLAS	4.5	pulsed	near ambient	TE	56	7×10^{-9}	N_2O	6.9×10^{7}	100
[39,68]	QCLAS	5.3	cw	near ambient	TE	210	3×10^{-10}	NO	2.0×10^{7}	30
[40,89]	QCLAS	5.3	pulsed	near ambient	cryogenic	210	[c] 5×10^{-10}	NO	1.9×10^{8}	1
					TE	210	[c] 1×10^{-9}	NO	4.1×10^{8}	1
[42,90]	PAS	7.9	cw	cryog./amb.	-	-	$\sim9\times10^{-7}$	CH_4	8.4×10^{11}	~150
								N_2O	3.5×10^{11}	~150
[43,91]	QEPAS[g]	4.6	cw	cryogenic	-	-	8×10^{-7}	N_2O	6.5×10^{9}	[b]
[41,92]	QEPAS[g]	8.4	ext. cav.	near ambient	-	-	4×10^{-7}	C_2HF_5	2.5×10^{9}	100

Table D.1: Intercomparison of high sensitivity QCL based spectrometers employing optical cavities, long path cells or PAS.

a) value estimated by the authors from the cited data
b) calculated from $n = p/(k_B T)$ multiplied by the best reported concentration or mixing ratio; normally $T = 296$ K
c) original value given in the reference
d) ICL
e) in combination with wavelength modulation the sensitivity was ~ 5 times better
f) converted into "per scan" value
g) quartz enhanced photoacoustic spectroscopy
h) 3 s time constant for a single spectral position

Bibliography

[1]   S.S. Brown, *Chem. Rev.* **103**, 5219 (2003).
[2]   E. Kerstel, L. Gianfrani, *Appl. Phys. B* **92**, 439 (2008).
[3]   M. Hori, T. Goto, *Plasma Sources Sci. Technol.* **15**, S74 (2006).
[4]   J. Ma, J. C. Richley, M. N. R. Ashfold, Y. A. Mankelevich, *J. Appl. Phys.* **104**, 103305 (2008).
[5]   S. Cheskis, *Prog. Energy Comb. Sci.* **25**, 233 (1999).
[6]   M.W. Todd, R.A. Provencal, T.G. Owano, B.A. Paldus, A. Kachanov, K.L. Vodopyanov, M. Hunter, S.L. Coy, J.I. Steinfeld, J.T. Arnold, *Appl. Phys. B* **75**, 367 (2002).
[7]   M.R. McCurdy, Y. Bakhirkin, G. Wysocki, R. Lewicki, F.K. Tittel, *J. Breath Res.* **1**, 014001 (2007).
[8]   M. Mürtz, P. Hering, *Phy. Journ.* **7** (10), 37 (2008) (*in German*)
[9]   A. O'Keefe, D.A.G. Deacon, *Rev. Sci. Instrum.* **59**, 2544 (1988).
[10] R. Engeln, G. Berden, R. Peeters, G. Meijer, *Rev. Sci. Instrum.* **69**, 3763 (1998).
[11] A. O'Keefe, J.J. Scherer, J.B. Paul, *Chem. Phys. Lett.* **307**, 343 (1999).
[12] A. O'Keefe, *Chem. Phys. Lett.* **293**, 331 (1998).
[13] M. Mürtz, B. Freech, W. Urban, *Appl. Phys. B* **68**, 243 (1999).
[14] J. Ye, L.S. Ma, J.L. Hall, *J. Opt. Soc. Am. B* **15**, 6 (1998)
[15] D. Romanini, A.A. Kachanov, F. Stoeckel, *Chem. Phys. Lett.* **270**, 538 (1997).
[16] G. Berden, R. Peeters, G. Meijer, *Chem. Phys. Lett.* **307**, 131 (1999).
[17] K.W. Busch, M.A. Busch (eds), *Cavity Ring-Down Spectroscopy: An Ultratrace Absorption Measurement Technique* (Amer. Chem. Soc., 1999).
[18] J.J. Scherer, D. Voelkel, D.J. Rakestraw, J.B. Paul, C.P. Collier, R.J. Saykally. A. O'Keefe, *Chem. Phys. Lett.* **245**, 273 (1995).
[19] J.J. Scherer, K.W. Aniolek, N.P. Cernansky, D.J. Rakestraw, *J. Chem. Phys.* **107**, 6196 (1997).
[20] S. Wu, P. Dupré, T.A. Miller, *Phys. Chem. Chem. Phys.* **8**, 1682 (2006).
[21] F. Ito, *J. Chem. Phys.* **124**, 054309 (2006).
[22] J.B. Paul, R.J. Saykally, *Anal. Chemistry* **69**, 287 A (1997).
[23] J.B. Paul, R.A. Provencal, C. Chapo, K. Roth, R. Casaes, R.J. Saykally, *J. Phys. Chem. A* **103**, 2972 (1999).
[24] J.B. Paul, R.A. Provencal, C. Chapo, A. Petterson, R.J. Saykally, *J. Chem. Phys.* **109**, 10201 (1998).
[25] R. Peeters, G. Berden, A. Olafsson, L.J.J. Laarhoven, G. Meijer, *Chem. Phys. Lett.* **337**, 231 (2001).
[26] D. Kleine, H. Dahnke, W. Urban, P. Hering, M. Mürtz, *Opt. Lett.* **25**, 1606 (2000).
[27] H. Dahnke, G. von Basum, K. Kleinermanns, P. Hering, M. Mürtz, *Appl. Phys. B* **75**, 311 (2002).
[28] D. Halmer, G. von Basum, P. Hering, M. Mürtz, *Opt. Lett.* **30**, 2314 (2005).
[29] K.W. Aniolek, P.E. Powers, T.J. Kulp, B.A. Richman , S.E. Bisson, *Chem. Phys. Lett.* **302**, 555 (1999).
[30] S. Stry, P. Hering, M. Mürtz, *Appl. Phys. B* **75**, 297 (2002).
[31] S.E. Bisson, T.J. Kulp, O. Levi, J.S. Harris, M.M. Fejer, *Appl. Phys. B* **85**, 199 (2006).

[32] R. Engeln, E. van den Berg, G. Meijer, L. Lin, G.M.H. Knippels, A.F.G. van der Meer, *Chem. Phys. Lett.* **269**, 293 (1997).

[33] R. Engeln, G. von Helden, A.J.A. van Roij, G. Meijer, *J. Chem. Phys.* **110**, 2732 (1999).

[34] M.M. Hemerik, Ph.D. Thesis, Eindhoven University of Technology, ISBN 90-386-1799-2, 2001.

[35] J. Remy, M.M. Hemerik, G.M.W. Kroesen, W.W. Stoffels, *IEEE Trans. on Plasma Science* **32**, 709 (2004).

[36] W.S. Tam, I. Leonov, Y. Xub, *Rev. Sci. Instrum.* **77**, 063117 (2006).

[37] D.D. Nelson, J.B. McManus, S. Urbanski, S. Herndon, M.S. Zahniser, *Spectrochim. Acta A* **60**, 3325 (2004).

[38] C. Roller, A.A. Kosterev, F.K. Tittel, K. Uehara, C. Gmachl, D.L. Sivco, *Opt. Lett.* **28**, 2052 (2003).

[39] D.D. Nelson, J.B. McManus, S.C. Herndon, J.H. Shorter, M. S. Zahniser, S. Blaser, L. Hvozdara, A. Muller, M. Giovannini, J. Faist, *Opt. Lett.* **31**, 2012 (2006).

[40] D.D. Nelson, J.H. Shorter, J.B. McManus, M.S. Zahniser, *Appl. Phys. B* **75**, 343 (2002).

[41] R. Lewicki, G. Wysocki, A.A. Kosterev, F.K. Tittel, *Opt. Express* **15**, 7357 (2007).

[42] A. Grossel, V. Zeninari, B. Parvitte, L. Joly, D. Courtois, *Appl. Phys. B* **88**, 483 (2007).

[43] A.A. Kosterev, Y.A. Bakhirkin, F.K. Tittel, *Appl. Phys. B* **80**, 133 (2005).

[44] A.A. Kosterev, R.F. Curl, F.K. Tittel, C. Gmachl, F. Capasso, D.L. Sivco, J.N. Baillargeon, A.L. Hutchinson, A.Y. Cho, *Appl. Opt.* **39**, 4425 (2000).

[45] B. Tuzson, M.J. Zeeman, M.S. Zahniser, L. Emmenegger, *Infrared Phys. Tech.* **51**, 198 (2008).

[46] J.B. McManus, P.L. Kebabian, M.S. Zahniser, *Appl. Opt.* **34**, 3336 (1995).

[47] B.A. Paldus, C.C. Harb, T.G. Spence, R.N. Zare, C. Gmachl, F. Capasso, D.L. Sivco, J.N. Baillargeon, A.L. Hutchinson, A.Y. Cho, *Opt. Lett.* **25**, 666 (2000).

[48] M.S. Taubman, T.L. Myers, B.D. Cannon, R.M. Williams, *Spectrochim. Acta Part A* **60**, 3457 (2004).

[49] M.L. Silva, D.M. Sonnenfroh, D.I. Rosen, M.G. Allen, A.O'Keefe, *Appl. Phys. B* **81**, 705 (2005).

[50] J. Manne, O. Sukhorukov, W. Jäger, J. Tulip, *Appl. Opt.* **45**, 9230 (2006).

[51] O. Sukhorukov, A. Lytkine, J. Manne, J. Tulip, W. Jäger, *Proc. SPIE* **6127**, 61270A (2006).

[52] L. Menzel, A.A. Kosterev, R.F. Curl, F.K. Tittel, C. Gmachl, F. Capasso, D.L. Sivco, J.N. Baillargon, A.L. Hutchinson, A.Y. Cho, W. Urban, *Appl. Phys. B* **72**, 859 (2001).

[53] J.B. Paul, J.J. Scherer, A. O'Keefe, L. Lapson, J.G. Anderson, C. Gmachl, F. Capasso, A.Y. Cho, *Proc. SPIE* **4577**, 1 (2002).

[54] Y.A. Bakhirkin, A.A. Kosterev, C. Roller, R.F. Curl, F.K. Tittel, *Appl. Opt.* **43**, 2257 (2004).

[55] A.A. Kosterev, A.L. Malinovsky, F.K. Tittel, C. Gmachl, F. Capasso, D.L. Sivco, J.N. Baillargeon, A.L. Hutchinson, A.Y. Cho, *Appl. Opt.* **40**, 5522 (2001).

[56] Y.A. Bakhirkin, A.A. Kosterev, R.F. Curl, F.K. Tittel, L. Hvodzdara, M. Giovannini, J. Faist, *Appl. Phys. B* **82**, 149 (2006).

[57] M.R. McCurdy, Y.A. Bakhirkin, F.K. Tittel, *Appl. Phys. B* **85**, 445 (2006).

[58] F.K. Tittel, Y. Bakhirkin, A.A. Kosterev, G. Wysocki, *Rev. Laser Eng.* **34**, 275 (2006).

[59] A.A. Kosterev, F.K. Tittel, *IEEE J. Quant. Electron.* **38**, 582 (2002).

[60] J.H. Miller, Y.A. Bakhirkin, T. Ajtai, F.K. Tittel, C.J. Hill, R.Q. Yang, *Appl. Phys. B* **85**, 391 (2006).

[61] E.J. Moyer, D.S. Sayres, G.S. Engel, J.M.St. Clair, F.N. Keutsch, N.T. Allen, J.H. Kroll, J.G. Anderson, *Appl. Phys. B* **92**, 467 (2008).

[62] M. Mazurenka, A.J. Orr-Ewing, R. Peverall, G.A.D. Ritchie, *Annu. Rep. Prog. Chem. Sect. C.* **101**, 100 (2005).

[63] P. Zalicki, R. N. Zare, *J. Chem. Phys.* **102**, 2708 (1995).

[64] L.S. Rothman *et al., J. Quant. Spectrosc. Radiat. Transfer* **96**, 139 (2005).

[65] The MDND was calculated with 50 ppb at atmospheric pressure. The *NEA* was estimated via the minimum round trip loss for 16 ppb standard deviation (figure 7 in [50]) and converted into k_{min}. Eq. (6 - 17) for 20 s averaging (without $\sqrt{2}$) yields the *NEA*.

[66] Figure 8 in [49] implies a minimum detectable absorption of 3×10^{-3} for 1500 m in 4 s. The *NEA* follows from eqs. (6 - 16) and (6 - 17) and the MDND from 0.7 ppb at 100 Torr or 0.8 ppb at 70 Torr. Those minimum concentrations are given as rms noise from ~ 10 different 4 s scans and are therefore smaller than a minimum absorption of 3×10^{-3} would suggest.

[67] Figure 3 in [52] implies an absorption of 0.0005 for 13 ppb NO (50 Torr) for a SNR = 5, i.e. the minimum detectable absorption is $\sim 10^{-4}$. For $N = 100$ scans the SNR was improved by $N^{1/2}$ while for 10000 averages an additional factor 2.5 is achieved, i.e. a factor 25 in total compared to a single scan. The *NEA* follows from eqs. (6 - 16) and (6 - 17) (without $\sqrt{2}$ because $L_{eff} = 100$ m was not calibrated).

[68] The Allan variance minimum of 0.03 ppb after 30 s corresponds to 3×10^{-6} absorbance noise and $k_{min} = 1.5 \times 10^{-10}$ cm^{-1} respectively [39]. Since the short term noise (1 s) is 0.06 ppb the corresponding values were also scaled by a factor 2, i.e. the *NEA* is $\sim 3 \times 10^{-10}$ cm^{-1}Hz$^{-1/2}$. The MDND follows from 0.03 ppb at 20 Torr.

[69] J.T. Hodges, J.P. Looney, R.D. van Zee, *Appl. Opt.* **35**, 4112 (1996).

[70] G. Berden, R. Peeters, G. Meijer, *Int. Rev. Phys. Chem.* **19**, 565 (2000).

[71] P. Macko, D. Romanini, N. Sadeghi, presented at the Frontiers in Low Temperature Plasma Diagnostics (FLTPD) IV, Rolduc, The Netherlands, 25–29 March 2001 (http://www.phys.tue.nl/FLTPD/poster/macko.pdf).

[72] I. Labazan, S. Rudic, S. Milosevic, *Chem. Phys. Lett.* **320**, 613 (2000).

[73] A value of $A_{ul} = 6.153$ s^{-1} is given in [64] for the strongest absorption feature in table 6.1 at 1306.929 cm^{-1}. The weighted transition dipole moment follows from (A - 5) to 0.094 Debye. A QCL output power of ~ 25 mW (sbcw848) is reported in its datasheet for the operating conditions used here. Additional beam shaping optics (e.g., ZnSe optics, OAP telescope) should reduce the available power to about 10 mW at the cavity entrance mirror. Imaging with an IR camera let us further assume a beam diameter of 10 mm. From detector signals measured with and without cavity we assume a fraction of 10^{-3} to be transmitted through the first mirror. Using (2 - 5) the electric field E_0 is then ~ 16 V/m inside the cavity. The relaxation rate γ in (2 - 27) was estimated for 2 mbar from $\gamma_{air,ref} = 0.0716$ cm^{-1}/atm [64] using (2 - 26).

[74] E. Uherek, R. Sander, *ESPERE Climate Encyclopaedia*, (http://www.espere.net/, 2006).

[75] P. Werle, R. Mücke, F. Slemr, *Appl. Phys. B* **57**, 131 (1993).

[76] D.W. Allan, *Proc. IEEE* **54**, 221 (1966).

[77] L_{eff} was calculated from $\tau_0 = 3.5$ µs. After 8 s the relative error of τ was 4.7×10^{-3}; eqs. (6 - 15) and (6 - 17) yield the *NEA* (without $\sqrt{2}$ because R was not calibrated). The MDND follows from the detection limit of 0.7 ppb at 60 Torr [55].

[78] After 600 scans at 600 Hz the standard deviation was 10^{-4}. The *NEA* follows from eqs.

(6 - 16) and (6 - 17) and the MDND from 0.25 ppb detection limit at STP [47]. L_{eff} was estimated from $\tau_0 = 0.93$ μs.

[79] The MDND was calculated with 16 ppb at 30 Torr [52]. A minimum detectable absorption of 0.01 with $L_{eff} = 670$ m in 200 s yields the NEA with eqs. (6 - 16) and (6 - 17).

[80] L_{eff} was calculated from $R_{eff} = 99.4$ % and 50 cm base length. The MDND follows from 100 ppb at 50 Torr whereas the NEA was calculated with 0.15 % standard deviation in absorption for 3.3 s averaging and $L_{eff} = 83.3$ m [60].

[81] Figure 10 in [54] implies a minimum detectable absorption of 3×10^{-4} for 75 m in 15 s. The NEA follows from eqs. (6 - 16) and (6 - 17) and the MDND from 10 ppb at 100 Torr.

[82] Figure 4 in [56] implies a minimum detectable absorption of 6×10^{-3} for 700 m in 1 s. The NEA follows from eqs. (6 - 16) and (6 - 17) and the MDND from 3.2 ppb at 200 Torr.

[83] Figure 5 in [57] implies a minimum detectable absorption of 6×10^{-4} for 500 m in 4 s. The NEA follows from eqs. (6 - 16) and (6 - 17) and the MDND from 3.6 ppb at 100 Torr.

[84] With the given gain factor 6760 R_{eff} is estimated to be 99.9852 %. Consequently L_{eff} is 5560 m for a base length of 82.3 cm. The MDND follows from the 3 ppb detection limit at 30 Torr [53].

[85] The MDND was calculated from the Allan variance minimum of 0.12 ppm at 14 Torr after 240 s corresponding to a peak absorbance precision of 6×10^{-5}. This value was scaled with ~ 10 (since it does not scale with the square root of averages [45]) to determine the short term deviation (1 s) yielding 1.1 ppm. The NEA follows from eqs. (6 - 16) and (6 - 17) for 56 m path length.

[86] The MDND was calculated for the smallest pressure given in [44] (20 Torr) and 2.5 ppb CH_4 and 1.0 ppb N_2O respectively. The NEA follows from eqs. (6 - 16) and (6 - 17) (without $\sqrt{2}$) for 3.5×10^{-5} minimum peak absorbance, 30 s averaging and 100 m path length.

[87] The NEA follows from eqs. (6 - 16) and (6 - 17) for 1.4×10^{-4} absorbance precision at 1 s sampling rate. For N_2O this yields a MDND calculated from 3 ppb at 60 Torr. For CH_4 the MDND was estimated from the Allan variance after 200 s: 0.7 ppb at 50 Torr [37].

[88] Figure 5 in [37] implies a minimum absorbance of 4×10^{-5} for a 1 Hz sampling rate at 56 m path length which yields the (short term) NEA from eqs. (6 - 16) and (6 - 17) (without $\sqrt{2}$). The MDND follows from 35.5 Torr (figure 3) and the Allan variance minimum of 0.06 ppb.

[89] The MDND was calculated for 48 Torr and 1 s averaging time (figure 3) by means of the given NEAs, i.e. 0.12 ppb and 0.26 ppb for the LN and TE cooled detector respectively [40].

[90] The MDNDs follow from 34 ppb and 14 ppb for CH_4 and N_2O at atmospheric pressure respectively. The NEA was estimated from the noise of 0.4 μV and the calibration factor of 715 V/(Wcm^{-1}) scaled with the output power of 8 mW [42].

[91] The MDND follows from 4 ppb at 50 Torr for a 3 s lock-in time constant. The normalised NEA was scaled with the output power of 19 mW [43].

[92] The MDND was calculated with 0.1 ppb at 770 Torr after 100 s (Allan variance plot). The normalised NEA was scaled with the output power of 6.6 mW [41].

7 Accuracy and limitations

The previous chapters concerned a few aspects of sensitivity (detection limit) and accuracy comprising both reproducibility and precision of the measurements. In what follows the main limiting factors in practise are generalised and discussed in terms of the spectroscopic application. Sometimes the achieved (and required) detection and accuracy limits differ substantially from well-established figure of merits. The discussion focuses on

i) TDLAS in reactive plasmas using laser sweep integration (chapter 3),
ii) time resolved QCLAS for pulsed plasmas (chapter 5),
iii) (atmospheric) trace gas measurements employing CEAS (chapter 6).

Generally, the sweep integration method implemented here by means of *TDL Wintel* provides an optimised sensitivity in direct absorption for a specific transition along with several convenient tools, e.g., online background subtraction for suppressing optical fringes and atmospheric background absorption (e.g., H_2O, CO_2), respectively. The relative uncertainty is directly linked to the selected time resolution, because all laser sweeps are accumulated and averaged until the so called data update time (i.e., time resolution) is reached. For measurements in almost constant gas mixtures, among them trace gas detection, a low time resolution and averaging up to the usually reported Allan minimum may be appropriate. Time resolved measurements in plasmas (i), however, require a higher time resolution with an inherently reduced SNR. The relative uncertainty of the detectors signal (half of the peak-to-peak noise level) was typically not better than 10^{-2} for 0.1 ... 1 s data update times where the baseline noise strongly depends on the quality of the detector power supply. The theoretical sensitivity limit is then determined by the number of passes of the multi-pass cell ($L_{eff} \leq 60$ m) and the line strength of the transition ($\sim 10^{-20}$ cm/molecule). Using equation (2 - 16) (low pressure case, $\Delta \nu_D \sim 10^{-3}$ cm$^{-1} \leq \Delta \nu_{obs}$) this yields a MDND $n_{min} \approx 3 \times 10^{11}$ cm^{-3}.

However, in practice the application to plasmas introduces several other and more important sources of uncertainty lowering both the reproducibility and the precision of the results. Since mercury cadmium telluride (MCT) detectors are also sensitive to radiation in the NIR and visible spectral range instabilities and fluctuations of the plasma may increase the noise level considerably. In many cases suppression by means of a second monochromator or filters is not an option due to the limited signal levels of lead salt lasers in combination with multiple pass arrangements. The final relative error of molecular number densities is typically within 10 % and may be deduced from individual plots of concentration against time.

A similar estimate for the precision uncertainty is even more difficult to obtain: The usually unknown gas temperature is a major parameter for determining absolute number densities and strongly influences the partition function $Q(T)$ (eqs. (2 - 19), (2 -20)) and thus the line strength. A post-measurement correction is required in *TDL Wintel*, which includes an approximation for $Q(T)$ [1] which is less accurate for higher temperatures and molecules formed by more than two atoms. Particularly for hydrocarbons the error may readily exceed 10 % as soon as the gas temperature increases above ~ 400 K which is normally the case in microwave plasmas. Presently, the gas temperature in the microwave reactor may vary by ± 20 % and so is the relative precision error (to a first order approximation). Transitions with high lower state energies E_l might exhibit an increased uncertainty.

Since direct absorption spectroscopy is a line of sight method, the applications of multi-pass cells yields an average or effective gas temperature (e.g., across 1.5 m base length in the planar microwave reactor). This could be overcome by means of CEAS methods yielding similar or even higher absorption paths at shorter base lengths and therefore more localised measurements. In some cases the comparison of simultaneously recorded lines (with different E_l) of the same molecule might help qualifying an upper and lower limit. A similar challenge is the determination of an effective path length L_{eff} for radical measurements, because these molecules may not be produced and be present across the entire reactor base length. If the same L_{eff} is used as for stable species, the radical concentration may systematically be underestimated by up to 25 % in the present experiments. This follows from the assumption that radicals are mainly present in the brightest part of the plasma [2].

The inherent multimode behaviour of tuneable lead salt lasers might reduce the precision further, especially in the case of closely spaced laser modes that cannot adequately be separated by a monochromator. Measurements are typically carried out with less than 5 % contribution from competing laser modes. Concentrations from weak and moderate absorption lines ($\leq$ 10 %) without additional broadband absorption are then underestimated by up to 5 %. In all other cases the multimode offset has to be carefully corrected or calibrated since $n \sim \ln(I_0/I)$ (eq. (2 -4)) causes a strong non-linear increase of the errors with decreasing transmission I.

Considering all limiting factors along with TDLAS measurements in plasmas it is clear that absolute number densities are validated within a factor of two which is usually good enough for further conclusions. In contrast, trace gas measurements would require a much better accuracy, particularly, a better precision which can be realised by establishing a well-defined gas temperature.

The accuracy of QCLAS and non-linear absorption effects inherent to short or long laser pulses have been extensively discussed in chapter 4. Below the uncertainties of time resolved measurements in plasmas employing the *intra* pulse mode and calibrated absolute number densities (ii) are summarised.

The required time resolution of an *intra* pulse study establishes the data acquisition mode and therefore the data treatment and error sources. If a resolution of 200 µs or less (< 5 kHz) is sufficient, scanning through the plasma pulse can be accomplished by exploiting the (internal) repetition rate of the pulsed QCL where the repetition frequency sets the time resolution [3,4]. The superposition of several full plasma scans facilitates spectral averaging and can conveniently be carried out post-measurement. Conversely, for higher time resolutions a delayed trigger scheme is necessary (chapter 5) where only one spectrum per plasma pulse is acquired. Averaging is accomplished over successive plasma pulses and a full scan of the pulsed discharge is rather time consuming. A specific problem of the latter method is that for each spectrum a valid trigger event (e.g., the slightly irreproducible current through the plasma) is necessary. In contrast, by exploiting the internal QCL repetition frequency a single scan across the entire plasma pulse consisting of several individual spectra is facilitated by only one trigger event.

Both methods require spectral averaging of minimum 20 single QCL pulses in order to sufficiently suppress the (vertical) pulse-to-pulse fluctuations ($\pm$ 15 %) and the (horizontal) jitter ($\pm$ 1 ns) between TTL and optical pulse of the QCL. Additionally, a long-term drift (within a few hours) of the QCL baseline (I_0) in its absolute value and shape is observed. Specifically, for this reason time consuming measurements with the delayed trigger scheme are challenging.

For discharge conditions where averaging is feasible, i.e. under flowing gas conditions, the discussed uncertainties typically result in a minimum detectable absorption (half of the peak to peak noise level) of 3×10^{-2} and a relative error of less than 10 % in the number densities. Applying equation (2 - 16) and figure 4.12 to estimate the chirp rate limited resolution ($\Delta \nu_{min} \sim 5 \times 10^{-3}$ cm^{-1}) the minimum noise level converts to a MDND $n_{min} \approx 6 \times 10^{14}$ cm^{-3} in single pass configuration ($L_{eff} = 50$ cm) for a line strength of $S = 10^{-20}$ cm/molecule. This pictures changes if static measurements (single shot) are performed, since then the detector noise level of a few mV in combination with the QCL output power and the final detector signal govern the relative concentration error of up to 20 %. The uncertainty for both averaged and unaveraged experiments include potential inaccuracies in determining the integration limits for calculating the integrated absorption coefficient. Hence, potential gain in time resolution by employing the *intra* pulse mode compared to TDLAS sweep integration methods is at the expense of the SNR or reproducibility. The systematic (precision) error due to power saturation or fast passage is balanced by the calibrated correction factor.

Nevertheless, the achieved accuracy was good enough to observe (relative) temperature induced changes in the integrated absorption coefficient in the same order of the mentioned uncertainty levels. Additional inaccuracies originating from plasma diagnostics (~ factor of 2) as discussed for TDLAS are also present here. Fortunately, DFB QCLs are completely free of multimode concerns. Note that chirp rate induced band width limitations (chapter 4) may also reduce the sensitivity in the *intra* pulse mode. Currently, this may inevitably occur for uncommon spectral positions at the edge of the MIR spectral range (< 4 μm, > 12 μm) due to the slow response time of the available detector elements rather than the subsequent amplifying or digitising electronics.

A different situation has to be considered for trace gas measurements in gas mixtures (iii) where temperature induced inaccuracies are usually absent or can be obviated. Apart from high selectivity and sensitivity concerns, which generally require low pressure conditions and increased path lengths, the precision of the absolute number densities is of importance. A relatively straightforward experimental setup was in the centre of interest of the presented CEAS approach. Therefore an unlocked optical resonator aligned on-axis was employed and tested which could readily be transferred to plasma reactors for providing *in-situ* diagnostics. The residual noise level in a 1 s interval is typically not much less than 10^{-2} (figure 6.19) due to incomplete averaging over the mode structure of the cavity. Using equations (2 - 16) and (6 - 10) the MDND is estimated to be $n_{min} \approx 8 \times 10^{10}$ cm^{-3} for $L_{eff} = 1$ km (i.e., 50 cm base length, 99.95 % mirror reflectivity) and $S = 10^{-20}$ cm/molecule. For trace gas measurements transitions having an order of magnitude higher line strength are usually chosen and the decrease of $S(T)$ at higher temperatures is absent.

Achieving even better sensitivity due to an increased SNR would be accomplished by:

a)   increasing the mirror separation and
       slightly dithering the cavity length or QCL current,
b)   increasing the laser sweep rate,
c)   off-axis alignment [5]
d)   optical feedback to the laser [6,7],
d)   frequency locked cavities [8,9 and reference therein],

Currently, by means of option (a) the sensitivity has already been increased by one order of magnitude. The result should be regarded as an optimum, although a denser mode pattern due to an extended cavity base length is later difficult to realise at a reactor with fixed dimensions of several ten cm. Increasing the laser sweep rate (b) provides a better averaging of spectra within the same time interval (e.g., 1 s), but is limited by two substantial criteria. Firstly, the sweep rate must be slow enough to avoid the internal frequency chirp of the QCL as known from pulsed lasers. Secondly, an adequately long overlap with the cavity modes is essential [10]. The sweep rates in the present experiments of ~ 1 kHz (i.e. one longitudinal mode scanned in three times the ring-down time) should be regarded as a maximum sweep rate. Sweeping the QCL more rapidly results in strongly asymmetric line shapes. On the other hand slower scanning requires longer averaging (> 1 s) for achieving a reasonable SNR. It is also evident that this highly sensitive and straightforward CEAS approach does not simultaneously provide a high time resolution (about 1 ms per unaveraged single spectrum). Off-axis alignment is a clever option (c), since it increases the number of transverse modes and L_{eff} simultaneously. However, a more complex optical arrangement is necessary for accomplishing small QCL beam diameters. Moreover, the alignment in the MIR range - even with forthcoming IR cameras - is difficult [11]. Optical feedback (as a special case of frequency locking) for enabling a better intensity build-up inside and a higher throughput of the cavity (d) has been successfully applied with near infrared diode lasers, but is more challenging for QCLs. Custom-made QCLs with anti-reflection coatings would be necessary which follows from experiments with standard (uncoated) lasers in external cavity configuration [12]. Additionally, V-shaped cavities are then suggested which would be difficult to realise for plasma diagnostic purposes and degrade the spatial resolution. Accidental feedback to the QCL and its influence on the SNR should be scrutinised in the future or suppressed by a $\lambda/4$ plate as a precaution. Finally, frequency locked cavities (d) could certainly increase the sensitivity (by up two orders of magnitude in combination with FM techniques). In this case a better stabilised QCL power supply, a reference feedback cavity and considerable additional equipment [13] are necessary which would be far beyond a straightforward setup.

The precision of the retrieved absolute number densities clearly depends on the accuracy of the calibration, in other words the calibration of the (effective) mirror reflectivity is essential. Preliminary CRDS experiments may not improve their accuracy, because other cavity modes could be excited in this experiments. A calibration with standardised gas mixtures, ideally with those species that are finally measured, is desirable. Note that a calibrated mirror reflectivity of 99.94 % or 99.96 % (if R is 99.95 %) already causes an error of 20 % in L_{eff} and thus in the concentration. For this reason care must be taken with the vacuum and gas supply system for trace gas measurements.

As discussed earlier, the requirements for the calibration (and precision) are slightly less stringent for plasma diagnostics, especially for radical detection. In this case a gas specific calibration is impossible in any case. Since all equations for the analysis of CEAS spectra are based on the weak absorption assumption the influence of non-linear effects on weak lines (chapter 6, cf. figure 6.14) superimposed on the wing of strong (e.g., precursor gas) lines has to be verified. Furthermore, a small inert gas flow for protecting the cavity mirrors from the plasma environment is suggested. First experiments demonstrated that a compromise between a sufficient gas flow rate and a quenching of the active plasma zone (i.e. of L_{eff}) has to be found. The calibration should then be carried out with a protecting gas flow similar to the (plasma) measurement conditions.

Bibliography

[1]   R.A. McClatchey, W.S. Benedict, S.A. Clough, D.E. Burch, R.F. Calfee, K. Fox, L.S. Rothman, J.S. Garing, *Environm. Res. Pap.* **434**, Air Force Cambridge Research Laboratories (AFCRL) Technical Report (TR) 73-0096 (1973).

[2]   F. Hempel, Ph.D. Thesis, University Greifswald, ISBN 978-3-8325-0262-1, 2003.

[3]   S. Welzel, S. Stepanov, J. Meichnser, J. Röpcke *Journ. Phys.: Conf. Series* **157**, 012010 (2009).

[4]   S. Stepanov, S. Welzel, J. Röpcke, J. Meichnser *Journ. Phys.: Conf. Series* **157**, 012008 (2009).

[5]   J.B. Paul, L. Lapson, J.G. Anderson *Appl. Opt.* **40** 4904 (2001)

[6]   J. Morville, S. Kassi, M. Chenevier, D. Romanini *Appl. Phys. B* **80** 1027 (2005).

[7]   R. Wehr, S. Kassi, D. Romanini, L. Gianfrani *Appl. Phys. B* **92** 459 (2008).

[8]   D. Romanini, A.A. Kachanov, N. Sadeghi, E Stoeckel *Chem. Phys. Lett.* **264** 316 (1997).

[9]   N.J. van Leeuwen, J.C. Diettrich, A.C. Wilson *Appl. Opt.* **42** 3670 (2003).

[10] R. Engeln, G. Berden, R. Peeters, G. Meijer *Rev. Sci. Instrum.* **69** 3763 (1998).

[11] Y. Xu, X. Liu, Z. Su, R.M. Kulkarni, W.S. Tam, C. Kang, I. Leonov, L. D'Agostino *Proc. SPIE* **7222** 722208 (2009).

[12] G.P. Luo, C. Peng, H.Q. Le, S.S. Pei, W.Y. Hwang, B. Ishaug, J. Um, J.N. Baillargeon, C.H. Lin *Appl. Phys. Lett.* **78** 2834 (2001).

[13] M.S. Taubman, T.L. Myers, B.D. Cannon, R.M. Williams *Spectroch. Acta A* **60** 3457 (2004).

8   Summary and outlook

8.1   Summary

Infrared laser absorption spectroscopy (IRLAS) has been known as a valuable plasma diagnostic tool, specifically in molecular plasmas, for many years. Advanced IRLAS techniques employing tuneable lead salt and quantum cascade lasers (TDLs, QCLs) are at the centre of interest of this thesis thereby focussing on low pressure plasmas and applications such as atmospheric trace gas measurements.

The main results obtained in the present work can be summarised as follows:

i)   Microwave discharges containing either $Ar/H_2/N_2/O_2$ or $Ar/CH_4/N_2/O_2$ have been studied by means of TDLAS. Strong evidence for surface dominated molecule conversion, especially for $Ar/H_2/N_2/O_2$ mixtures, were found for the planar microwave reactor which so far has been described only by pure gas phase kinetics. The generalised Yasuda (reactor) parameter was found to be inapplicable for the parameter range used here, which was adjusted to the experimental conditions in a remote expanding thermal plasma (ETP). Although the reactor wall materials were significantly different, a comparison revealed the same trends for NH_3, N_2O and NO formation in the H_2 - N_2 - O_2 system.
The hydroxyl radical has been measured for the first time *in-situ* in a low temperature, low pressure plasma using conventional long path cell TDLAS. OH number densities ranged between 10^{11} ... 10^{12} cm^{-3} in both plasma chemical systems and reflected mainly the active plasma zone concentration.
By analysing the trends in 12 other measured molecular species (CH_4, C_2H_2, C_2H_4, C_2H_6, CO, CO_2, H_2O, H_2CO, NH_3, HCN, NO, N_2O) in $Ar/CH_4/N_2/O_2$ discharges a transition between deposition and etching conditions at $CH_4/O_2 \sim 1:1$, and an incomplete oxidation process of the precursor in general, were observed.

ii)   The application of QCLs for IRLAS under low pressure conditions employing the most common laser tuning approaches has been investigated in detail. Due to the considerably high input power deposited inside the QCL structure these new type of tuneable MIR light sources exhibit a frequency down chirp during pulsed operation. Average chirp rates were in the order of 0.005 cm^{-1}/ns (150 MHz/ns). The actual value varies strongly between different devices and also during the pulse-on time.
Both the high chirp rate and output power levels of several tens of mW of QCLs lead to non-linear absorption effects in fundamental ro-vibrational transitions under low pressure conditions, among them fast passage effects and power saturation of strong transitions which are the ones usually preferred for practical purposes in IRLAS. A new method of analysing absorption features quantitatively without preliminary calibration, even though their line shape is disturbed by the rapid passage effect, is proposed. If power saturation is negligible, integrating the undisturbed half of the line profile yields accurate calibration free number densities.

A time resolved analysis of individual QCL pulses in the *inter* pulse mode, which was adapted from conventional TDLAS, revealed that the QCL frequency chirp, the applied current tuning ramp and bandwidth limitations of the detection system cause an artificially broadened (effective) laser line width and asymmetric absorption line shapes. A detailed analysis of reported *inter* pulse spectrometers demonstrated that this method is generally not calibration free and affected by non-linear absorption phenomena (fast passage, power saturation) which are usually hidden by the limited detection bandwidth. An estimate of the minimum analogue bandwidth of such spectrometers is provided which should normally not fall below 250 MHz.

iii)   QCLAS using pulsed lasers has been used for time resolved plasma diagnostics for the first time enabling a time resolution down to about 100 ns to be achieved. The temperature evolution and heavy species kinetics in pulsed DC discharges containing Ar, N_2 and traces of NO have been investigated. A temperature increase of typically less than 50 K has been established for moderate current values in these discharges. This was achieved by comparing the time evolution of the NO concentration under static and flowing conditions and simplified model calculations. The relatively small gas heating during short pulses has a strong influence on the line strength of the NO absorption line and makes the line strength vary while the concentration remains almost constant during several hundred μs.

A recently developed model for such pulsed DC discharges containing N_2 - O_2 provided a means for identifying the main NO production and depletion rates. While $NO(X) + N_2(A) \rightarrow NO(A) + N_2$ is balanced by $NO(A) \rightarrow NO(X) + hf$ throughout the entire plasma pulse (ranging from 1 to 100 ms) the main loss channels are $NO(X) + N$ and $NO(X) + N(^2D)$. The onset of a significant decrease of NO is observed between 5 and 10 ms.

iv)   An alternative approach to conventional linear absorption spectroscopy employing multiple pass cells for achieving high sensitivity is to combine a high finesse cavity with thermoelectrically (TE) cooled QCLs and detectors. The sensitivity limits of an entirely TE cooled system equipped with a ~ 0.5 m (1.3 m) long cavity having a small sample volume of 0.3 l (1.2 l). With this spectrometer cavity enhanced absorption spectroscopy (CEAS) employing a continuous wave QCL yielded path lengths of 1080 m (4 km) and a noise equivalent absorption down to 4×10^{-8} cm^{-1}Hz$^{-1/2}$. The molecular concentration detection limit with a 20 s integration time was found to be 6×10^8 cm^{-3} for N_2O and 2×10^9 cm^{-3} for CH_4 at (1307 cm^{-1} or 7.66 μm) which is good enough for the selective measurement of trace atmospheric constituents at 2.2 mbar employing the short cavity. The detection limits of CH_4, N_2O and C_2H_2 for the 1.3 m long cavity were estimated to be generally below 1×10^{10} cm^{-3} for 1 s integrations time and one order of magnitude less for 30 s integration time (Allan minimum at 90 s). The main limiting factor for achieving even higher sensitivity, such as that found for larger volume multi pass cell spectrometers, is the residual mode noise of the cavity.

On the other hand the application of TE cooled pulsed QCLs for integrated cavity output spectroscopy and cavity ring down spectroscopy (CRDS) was found to be limited by the intrinsic frequency chirp of the laser. An efficient intensity build-up inside the cavity is suppressed. Consequently the accuracy and advantage of an absolute internal absorption calibration, in theory inherent for CRDS experiments, is not achievable. The full chirp of a QCL pulse rather than the effective laser line width was found to be the critical parameter in order to estimate potential bandwidth effects in a combination of pulsed QCLs with optical resonators.

8.2   Outlook

While the feasibility of highly sensitive chemical sensing by combining (continuous wave) QCLs with optical resonators has been established and optimised for low pressure trace gas measurements in this thesis, CEAS in the MIR spectral range may now also provide a powerful means in plasma diagnostics. Effective path lengths comparable to the 200 m limit, usually available from astigmatic Herriott type multiple pass cells, and beyond this limit can be accomplished *in-situ* by coupling an optical cavity to a plasma reactor. Figure 8.1 shows an first example of NO detected by CEAS in an Ar/O_2 planar microwave reactor where nitrogen was only present from a small leak in the reactor. The cavity having a base length of only 20 cm was installed perpendicular to the conventional White type multi pass cell. The estimated absorption path is about 500 m and thus suggests

    i)   an increase in sensitivity in *in-situ* plasma diagnostics, i.e. decreasing the minimum detectable number density, especially for highly reactive species, and

    ii)   localised measurements in a reduced plasma volume with effective absorption paths being sufficiently long for achieving reasonable detection limits.

It should be mentioned that both absorption features in figure 8.1 consist of unresolved transitions at 1818.66 cm^{-1} and 1818.78 cm^{-1} having a significantly different lower level energy. The measured transmission spectrum can thus be understood as a superposition of contributions from (hot) NO in the active zone and (colder) NO in the background part across the 20 cm cavity base length.

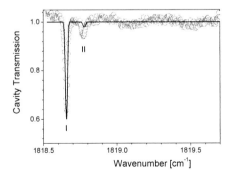

Figure 8.1:
CEAS spectrum of NO (feature I and II) observed in an Ar/O_2 microwave plasma (1.5 mbar, 1.5 kW, N_2 present from a small leak). Experimental data (circles) was fitted (solid black) by assuming a constant gas temperature across the cavity base length (20 cm). The mismatch for line II having an increased lower level energy indicates a non-uniform temperature distribution.

    Forthcoming widely tuneable QCL sources (external cavity lasers) may also facilitate the detection of broadband absorption features, e.g., of organic molecules or higher hydrocarbons, in atmospherics science, medical diagnostics (breath gas analysis) or plasma diagnostics.

The recent advent of QCLs has made possible the extension of IRLAS instrumentation to either higher sensitivity or higher time resolution. The selective application of these advanced techniques should support further discrimination between surface or gas phase related molecule conversion processes in reactive plasmas. While a higher time resolution of QCLAS spectrometers in combination with time dependent model calculations enables a better understanding of gas phase reaction kinetics, the increase in sensitivity in *in-situ* diagnostics may help in detecting highly reactive or sticky radicals, such as C_2H in the future.

9   Glossary

9.1   Acronyms

BW	bandwidth
cw	continuous wave
CALOS	cavity leak-out spectroscopy
CEA(S)	cavity enhanced absorption (spectroscopy)
CRD(S)	cavity ring down (spectroscopy)
DAC	data acquisition card
DC	direct current
DFB	distributes feedback
DFG	difference frequency generator
EC(L)	external cavity (laser)
EEDF	electron energy distribution function
ETP	expanding thermal plasma
FEL	free electron laser
FM	frequency modulation
FSR	free spectral range
FTS	Fourier transform spectrometer
FT-IR	Fourier transform infrared (spectroscopy)
FWHM	full width at half maximum
GC	gas chromatography
GEISA	gestion et étude des information spectroscopiques atmosphérique
HITRAN	high resolution transmission
HWHM	half width at maximum
ICL	interband cascade laser
ICOS	integrated cavity output spectroscopy
IRLAS	infrared laser absorption spectroscopy
LIF	laser induced fluorescence
LN	liquid nitrogen
MCT	mercury cadmium telluride
MDND	minimum detectable number density
MIR	mid infrared
MQW	multi quantum well
MS	mass spectrometry
MW	microwave
NEA	noise equivalent absorption
NICE-OHMS	noise-immune cavity enhanced optical heterodyne molecular spectroscopy
NIR	near infrared
NIST	National Institute of Standards and Technology
OAS	optical absorption spectroscopy
OAP	off-axis parabolic
OES	optical emission spectroscopy
OPO	optical parametric oscillator

PAS	photoacoustic spectroscopy
PNNL	Pacific Northwest National Laboratory
PP	periodically poled
ppb	parts per billion (10^{-9})
ppm	parts per million (10^{-6})
QCL	quantum cascade laser
QCLAS	quantum cascade laser absorption spectroscopy
QEPAS	quartz-enhanced photoacoustic spectroscopy
QPM	quasi phase matching
QW	quantum well
RF	radio frequency
sccm	standard cubic centimetre
SI	Système International d'Unités
SL	superlattice
SNR	signal-to-noise ratio
STP	standard temperature and pressure
TDL	tuneable diode laser
TDLAS	tuneable diode laser absorption spectroscopy
TE	thermoelectrical
TEC	thermoelectric cooler
TTL	transistor-transistor logic
UHV	ultra-high vacuum
UV	ultraviolett
VUV	vacuum ultraviolett
WM	wavelength modulation

9.2 Symbols

$\cdots_{ref}$	reference value
A	normalised sweep rate
α	sweep rate
A_{ul}	Einstein coefficient of spontaneous emission
B_{lu}, B_{ul}	Einstein coefficients of induced absorption and emission
B_c	carbon balance
BW	(analogue) bandwidth
c	velocity of light
d	single, one-dimensional extension of a medium in a cavity
D	precursor depletion
ε_0	dielectric constant
$E\ (E_0)$	electric field
E_l	lower state energy
E_{gap}	band-gap energy
ΔE_{disc}	band-gap discontinuity
E_{mean}	mean energy per feed gas molecule
f	frequency ($\neq$ angular frequency)

f_0, f_{ul}	transition frequency
Δf	line width (in Hz)
Δf_{col}	homogeneous line width (in Hz)
Δf_D	inhomogeneous (Doppler) line width (in Hz)
df/dt	chirp rate
ϕ	absorption line profile
F	gas flow rate (volume/time interval)
g_l, g_u	statistical weight of the lower and upper state
$\gamma, (\gamma_l, \gamma_u)$	relaxation rate (of the lower and upper state)
$\gamma_p, \gamma_{air}, \gamma_{self}$	pressure, air, self broadening coefficient
G	gain
Γ	reactor parameter
h	Planck constant
I	(transmitted) intensity
I_0	(transmitted) intensity without absorption
I_{abs}	absorbed intensity
I_a	isotopic abundance
j	current density
k_B	Boltzmann constant
k	absorption coefficient
k_{min}	minimum detectable absorption coefficient
$\kappa = 2\pi/\lambda$	wavenumber in phase space
K	integrated absorption coefficient
λ	wavelength
λ_B	Bragg wavelength
L	cavity base length
L_{eff}	effective absorption path length
N_A	Avogadro number
m_{molec}	molecular mass
μ_0	magnetic constant
$|\mu_{lu}|$	transition dipole moment
M_{mol}	molar mass
n	number density
n_{min}	minimum detectable number density
n_l, n_u	number density in the lower and upper state of an (open) two-level system
n_e	electron density
n_{eff}	refraction index
$\nu = 1/\lambda$	wavenumber (in spectroscopy)
$d\nu/dt$	chirp rate
$\Delta\nu/dt_{100}$	average chirp rate
NEA	noise equivalent absorption
$\Delta\nu$	line width (in wavenumbers)
$\Delta\nu_{col}$	homogeneous line width (in wavenumbers)
$\Delta\nu_D$	inhomogeneous (Doppler) line width (in wavenumbers)
Ω_{lu}	Rabi frequency
p	pressure

P_{in}	injected power
Q	total internal partition function
R	(effective) mirror reflectivity
R_f	fragmentation rate
S	line strength
SNR	signal-to-noise ratio
Σ	saturation parameter
Σ_{eff}	effective saturation parameter
σ	absorption cross section
σ_A	Allan variance
t	time
t_{on}	pulse-on time
t_{del}	delay time
t_{pulse}	time scale of an individual QCL pulse
τ	cavity decay constant (chapter 6)
	relaxation time (chapter 2)
τ_0	decay constant of an empty cavity
τ_{res}	residence time
τ_{heat}	characteristic heating time
T_1, T_2	longitudinal and transversal relaxation time
$T\,(T_g)$	(gas) temperature
T_e	electron temperature
U	fractional absorption
U_{QCL}	QCL voltage
V	volume

Danksagung/Acknowledgement

Da das Zustandekommen einer solchen Arbeit nicht ohne tatkräftige Mithilfe vieler Personen möglich ist, möchte ich einige davon im Folgenden erwähnen.

Ganz sicher unmöglich wäre diese Arbeit ohne meinen Mentor, Jürgen Röpcke, der zeitig die Initiative ergriff und mich in das Plasmadiagnostik-Team des INP Greifswalds "lotste". Für die sich damit eröffnenden Möglichkeiten, die eingeräumten Freiheiten in den vergangenen 4 Jahren und den kontinuierlichen Meinungsaustausch, insbesondere zum Ende der Arbeit, möchte ich mich hiermit bedanken.

Anschließend sei all jenen fleißigen Leuten gedankt, die normalerweise an solchen Stellen zu kurz kommen, ohne die aber nie etwas funktionieren würde. Die hier zusammengestellten Ergebnisse wären nicht ohne die permanente Unterstützung von Stephan, Frank (LKW), Uwe und Dietmar sowie dem "Einkauf" des INP, insbesondere I. von Rekowski und A. Wolters, möglich gewesen, die - trotz der recht zahlreichen durch mich verursachten Messreisen bzw. Zollformalitäten - immer unterstützend zur Seite standen.

Ein besonderer Dank gilt dem inoffiziellen Betreuer, Paul Davies. I remember that you mentioned once: "As you know the Anglo Saxons don't waste words!" This is obviously true, but your comments and questions were always to the point and provided much more guidelines than suggestions from any other person. Many thanks for not many, but very fruitful discussions and editing or proof-reading of numerous manuscripts including this one.

Bedingt durch die enge Zusammenarbeit mit der Arbeitsgruppe von Antoine Rousseau und Olivier Guaitella an der Ecole Polytechnique wurden mir zahlreiche Aufenthalte nahe Paris beschert. Ma connaissance de la langue française n'est pas encore suffisante d'exprimer ma gratitude. Therefore I would like to thank for the hospitality and a lot of (not only) scientific discussions which we can hopefully extend in the future. Ähnliches lässt sich für die Verbindungen zur Technischen Universität Eindhoven sagen: bereits nach wenigen Tagen in Greifswald konnte ich Daan Schram, Richard Engeln und Rens Zijlmans kennenlernen. Darauf folgten etliche gemeinsame Messungen an beiden Standorten. I really enjoyed this very productive collaboration. Particularly I am indebted to the initial support concerning CRDS/CEAS and for providing the mirror mounts.

Sehr hilfreich war die Zusammenarbeit mit Onno Gabriel während der ersten Monate am INP. Eine bessere Einführung zu effizienten IR-spektroskopischen Messungen kann man sich nicht wünschen. Bezüglich sämtlicher Fragestellungen rund um QCLs, deren vielleicht auch nicht ganz standardmäßiger Handhabung und Anwendung war Sven Glitsch ein grundsätzlich konstruktiver und unschätzbarer Wegbegleiter. Nicht nur rein fachlich, sondern auch im Hinblick auf die vielen nüchternen Ansichten wird mir zukünftig etwas fehlen (Die Message, Trainer :-))

Dem restlichen Plasmadiagnostik-Team und allen anderen, noch nicht namentlich genannten Kollegen inner- und außerhalb des INP danke ich für die vielfältige Unterstützung.

Der größte Dank geht an meine Familie. Ohne die Unterstützung meiner Eltern wäre kein Studium und damit die Voraussetzung für eine Promotion möglich gewesen. Auch wenn vielleicht nicht alles verständlich war in den letzten Jahren, haben sie und mein Bruder doch immer wieder für Ablenkung, hilfreiche Ratschläge und die nötige Bodenhaftung gesorgt.

Curriculum Vitae

Personal details

Stefan Welzel

Date of birth	6 May 1979
Place of birth	Meerane
Citizenship	German

Education

1985 - 1987	Juri-Gagarin-Oberschule, Glauchau
1987 - 1992	Pestalozzi-Oberschule, Glauchau
1992 - 1997	Georgius-Agricola-Gymnasium, Glauchau Qualification: Abitur (Level: 1.0)
1999 - 2004	Student at the Chemnitz University of Technology Qualification: Dipl.-Phys. (Level: 1.1) (Oct. 2005: awarded with the "Universitätspreis")
2005 onwards	Ph.D. student at the INP Greifswald

Experience

1998 - 1999	Compulsory Civilian Service (Stadtverwaltung Glauchau)
2002 & 2003	Internship (one month) at the INP Greifswald
2005 - 2009	Researcher at the INP Greifswald Student of the International Max-Planck Research School (IMPRS) Collaborative Research Centre (SFB-TRR 24)
2005 - 2008	Continuous collaborative (exchange) program with - Ecole Polytechnique, Palaiseau (France) - Eindhoven University of Technology (The Netherlands)

External measurement campaigns ($\geq$ 2 weeks)

2005, Jun.	Eindhoven University of Technology
2005, Aug.	Humboldt University Berlin
2007, May	Eindhoven University of Technology
2008, Feb. - Jun.	E.-M.-Arndt University Greifswald

Related Publications

List of Publications

S. Welzel, G. Lombardi, P. Davies, R. Engeln, D. Schram, J. Röpcke
Using quantum cascade lasers with resonant optical cavities as a diagnostic tool
Journal of Physics: Conference Series **157**, 012009 (2009).

S. Welzel, S. Stepanov, J. Meichnser, J. Röpcke
Application of quantum cascade laser absorption spectroscopy to studies of fluorocarbon molecules
Journal of Physics: Conference Series **157**, 012010 (2009).

S. Stepanov, St. Welzel, J. Röpcke, J. Meichnser
Time-resolved QCLAS measurements in pulsed cc-rf CF_4/H_2 plasmas
Journal of Physics: Conference Series **157**, 012008 (2009).

S. Welzel, G. Lombardi, P. B. Davies, R. Engeln, D. C. Schram, J. Röpcke
Trace gas measurements using optically resonant cavities and quantum cascade lasers operating at room temperature
J. Appl. Phys. **104**, 093115 (2008).

D. C. Schram, R.A.B. Zijlmans, J.H. van Helden, O. Gabriel, G. Yagci, S. Welzel, J. Röpcke, R. Engeln
Plasma processing and the importance of surface molecule generation
Journ. Optoelectronics Adv. Mat. **10** (8), 1904 (2008).

J. Röpcke, S. Welzel, N. Lang, F. Hempel, L .Gatilova, O. Guaitella, A. Rousseau, P. B. Davies
Diagnostic Studies of Molecular Plasmas Using Mid-Infrared Semiconductor Lasers
Appl. Phys. B **92**, 335-341 (2008).

S. Welzel, A. Rousseau, P.B. Davies, J. Röpcke
Kinetic and Diagnostic Studies of Molecular Plasmas Using Laser Absorption Techniques
Journal of Physics: Conference Series **86**, 012012 (2007).

S. Welzel, L. Gatilova, J. Röpcke, A. Rousseau
Time resolved study of NO destruction in a pulsed DC discharge using quantum cascade laser absorption spectroscopy
Plasma Sources Sci. Technol. **16**, 822 (2007).

R.A.B Zijlmans, O. Gabriel, S. Welzel, F. Hempel, J. Röpcke, R. Engeln, D.C. Schram
Molecule synthesis in an $Ar-CH_4-O_2-N_2$ microwave plasma
Plasma Sources Sci. Technol. **15**, 564 (2006).

Contributions

Oral Presentations

S. Welzel, O. Gabriel, R. Zijlmans, F. Hempel, R. Engeln, D.C. Schram, J. Röpcke
On measurements of the OH radical in planar microwave discharges by tunable diode laser absorption spectroscopy in the mid infrared
7th Colloquium Atmospheric Spectroscopy Applications, Reims, France, 6 - 8 September 2005

S. Welzel, G. Lombardi, R. Engeln, S. Glitsch, D.C. Schram, P.B. Davies, J. Röpcke,
Cavity ring down spectroscopy in the mid-infrared using pulsed quantum cascade lasers
5th CRD User Meeting, Oxford, Great Britain, 21 - 22 September 2005

S. Welzel, F. Hempel, S. Glitsch, R. Engeln, D.C. Schram, P.B. Davies, J. Röpcke
New developments in IR absorption spectroscopy of plasma processes
Technological Plasmas Workshop, Edinburgh, Great Britain, 8 - 9 December 2005

St. Welzel, M. Baudach
Untersuchungen zur Zersetzung von Kohlenwasserstoffen im Plasmagenerator PSI-2
XIII. Erfahrungsaustausch "Oberflächentechnologie mit Plasma- und Ionenstrahlprozessen", Mühlleithen, 14 - 16 March 2006

St. Welzel, G. Lombardi, R. Engeln, F. Hempel, S. Glitsch, D.C. Schram, P.B. Davies, J. Röpcke
Pulsed Quantum Cascade Lasers as a Light Source for Cavity Ring-Down Spectroscopy
DPG Frühjahrstagung (ISSN 0420-0195) 41 VI 5/2006, Augsburg, 27 - 30 March 2006

St. Welzel, L. Gatilova, P.B. Davies, R. Engeln, A. Rousseau, J. Röpcke
New Perspectives and Challenges Using Quantum Cascade Lasers for Plasma Diagnostics
XVIII ESCAMPIG (ISBN 2-914771-38-X), Lecce, Italy, 12 - 16 July 2006

St. Welzel, P.B. Davies, R. Engeln, J. Röpcke
Mid Infrared Absorption Spectroscopy Using Optical Cavities and Quantum Cascade Lasers
DPG Frühjahrstagung (ISSN 0420-0195) 42 VI 3/2007, Düsseldorf, 19 - 23 March 2007

St. Welzel, L. Gatilova, A. Rousseau, J. Röpcke
Time resolved absorption spectroscopic studies on the kinetics of a pulsed DC discharge using quantum cascade lasers
7th Workshop on Frontiers in Low Temperature Plasma Diagnostics (FLTPD), Beverley, Great Britain, April 01 - 05, 2007

St. Welzel, S. Glitsch, M. Hübner, J. Röpcke
Frontiers in quantum cascade laser absorption spectroscopy using different tuning mechanisms
2nd Workshop on Infrared Plasma Spectroscopy (IPS), Greifswald, 25 - 27 July 2007

St. Welzel, P.B. Davies, J. Röpcke
Optical cavity based absorption spectroscopy using quantum cascade lasers
7th CRD User Meeting, Greifswald, Germany, 18 - 19 September 2007

St. Welzel, G. Lombardi, P. Davies, R. Engeln, D. Schram, J. Röpcke
Using quantum cascade lasers with resonant optical cavities as a diagnostic tool
3rd Workshop on Infrared Plasma Spectroscopy (IPS), Greifswald, 23 - 25 July 2008

St. Welzel, P.B. Davies, R. Engeln, J. Röpcke
Cavity enhanced absorption spectroscopy using room temperature quantum cascade lasers
European Geosciences Union (EGU) General Assembly 2009, Vienna, 19 - 24 April 2009

Contributions
(only as first author, 28 other as co-author)

St. Welzel, M. Hübner, L. Gatilova, J. Röpcke
Side effects in using quantum cascade lasers for absorption spectroscopy
1st Workshop on Infrared Plasma Spectroscopy (IPS), Greifswald, 14 - 16 June 2006

St. Welzel, P.B. Davies, R. Engeln, G. Lombardi, J. Röpcke, D.C. Schram
Challenges in Using Quantum Cascade Lasers as a Light Source for CRDS
6th CRD User Meeting, Cork, Ireland, 18 - 19 September 2006

St. Welzel, O. Werhahn, C. Mann, J. Röpcke
On the accuracy of gas concentration measurements using pulsed quantum cascade lasers
226. PTB Seminars "Rückführbarkeit in der spektrometrischen Gasanalytik", Braunschweig
(PTB), 22 - 23 November 2006

St. Welzel, A. Rousseau, P.B. Davies, J. Röpcke
On recent progress in studying kinetics in molecular plasmas using laser absorption techniques
5th EU-Japan Joint Symposium on Plasma Processing, Belgrade (Serbia), 07 - 09 March 2007

St. Welzel, L. Gatilova, J. Röpcke, A. Rousseau
Time resolved absorption spectroscopic studies on the kinetics of a pulsed DC discharge using quantum cascade
lasers: dynamic of the gas heating
16th Int. Coll. on Plasma Processes (CIP), Toulouse, France, 04 - 08 June 2007

St. Welzel, L. Gatilova, O. Guaitella, C. Lazzaroni, A. Rousseau, J. Röpcke
Time-resolved quantum cascade laser absorption spectroscopy on NO containing pulsed DC discharges
2nd Workshop on Infrared Plasma Spectroscopy (IPS), Greifswald, 25 - 27 July 2007

St. Welzel , L. Gatilova , A. Rousseau , J. Röpcke
Time resolved absorption spectroscopic studies on the kinetics of a pulsed DC discharge using quantum cascade
lasers
Field Laser Applications in Industry and Research (FLAIR) 2007, Florence, Italy, 02 - 07
September 2007

St. Welzel, L. Gatilova, O. Guaitella, A. Rousseau, P. B. Davies, J Röpcke
Infrared Spectroscopic Studies of NOx Kinetics and Volatile Organic Compound Removal in Non-Thermal
Plasmas
6th Int. Symp. on Non-Thermal Plasma Technology (ISNTPT), Taipei, Taiwan, 12 - 16 Mai, 2008

St. Welzel, O. Guaitella, C. Lazzaroni, L. Gatilova, A. Rousseau, J. Röpcke
Time-resolved QCLAS measurements on nitric oxide containing DC discharges
XIX ESCAMPIG (ISBN 2-914771-04-5), Granada, Spain, 15 - 19 July 2008

St. Welzel, G. Lombardi, P.B. Davies, R. Engeln, D. Schram, J. Röpcke
Cavity enhanced absorption spectroscopy as a diagnostic tool based on quantum cascade lasers
XIX ESCAMPIG (ISBN 2-914771-04-5), Granada, Spain, 15 - 19 July 2008

St. Welzel, O. Guaitella, C. Lazzaroni, L. Gatilova,A. Rousseau, J. Röpcke
Time resolved studies on pulsed DC discharges using QCL
3rd Workshop on Infrared Plasma Spectroscopy (IPS), Greifswald, 23 - 25 July 2008

St. Welzel, O. Werhahn, J. Koelliker Delgado, D. Schiel, Ch. Mann, J. Wagner, J. Röpcke
On the accuracy of calibration free gas concentration measurements using pulsed QCL
3rd Workshop on Infrared Plasma Spectroscopy (IPS), Greifswald, 23 - 25 July 2008

St. Welzel, S. Stepanov, J. Meichnser, J. Röpcke
Application of quantum cascade laser absorption spectroscopy to studies of fluorocarbon molecules
3rd Workshop on Infrared Plasma Spectroscopy (IPS), Greifswald, 23 - 25 July 2008

St. Welzel, S. Stepanov, J. Meichnser, J. Röpcke
Time resolved QCLAS studies on pulsed fluorocarbon RF discharges
14. Fachtagung Plasmatechnologie (PT14), Wuppertal, 02 - 04 March 2009

St. Welzel , S. Stepanov, J. Meichsner , J. Röpcke
Time resolved studies on pulsed fluorocarbon plasmas using pulsed QCL
8th Workshop on Frontiers in Low Temperature Plasma Diagnostics (FLTPD), Blansko, Czech Republic, 19 - 23 April 2009

Relevant contributions as co-author

L. Gatilova, Y. Ionikh, S. Welzel, J. Röpcke, A. Rousseau
NO$_x$ production in a pulsed low pressure discharge in air : interaction with a porous semionductor surface
XXVIIth International Conference on Phenomena in Ionised Gases (ICPIG '05), Eindhoven, The Netherlands, 17 - 22 July 2005 (ISBN 90-386-2231-1)

R.A.B. Zijlmans, O. Gabriel, S. Welzel, G.D. Stancu, F. Hempel, J. Röpcke, R. Engeln, D.C. Schram
A different view on molecule formation in a CH$_4$ containing microwave plasma
XXVIIth International Conference on Phenomena in Ionised Gases (ICPIG '05), Eindhoven, The Netherlands, 17 - 22 July 2005 (ISBN 90-386-2231-1)

R.A.B. Zijlmans, O. Gabriel, S. Welzel, F. Hempel, J. Röpcke, R. Engeln, D.C. Schram
Molecule Synthesis in an Ar-CH4-O2-N2 Microwave Plasma
XVIII ESCAMPIG (ISBN 2-914771-38-X), Lecce, Italy, 12 - 16 July 2006

J.H. van Helden, O. Gabriel, R. Zijlmans, S. Welzel, G. Lombardi, G.D. Stancu, J. Röpcke, D.C. Schram, R. Engeln
Study of the Molecule Formation and Surface Coverage of the Reactor Wall in Ar/N$_2$/O$_2$ Plasmas
XVIII ESCAMPIG (ISBN 2-914771-38-X), Lecce, Italy, 12 - 16 July 2006

L. Gatilova, K. Allegraud, Y. Ionikh, S. Welzel, J. Röpcke, G. Cartry, A. Rousseau
NO Production During a Single Plasma Pulse in a Low Pressure Discharge
XVIII ESCAMPIG (ISBN 2-914771-38-X), Lecce, Italy, 12 - 16 July 2006

D.C. Schram, R.A.B. Zijlmans, J.H. van Helden, O. Gabriel, S. Welzel, J. Röpcke, R. Engeln
Molecular formation in plasmas and the role of surface processes
XIX ESCAMPIG (ISBN 2-914771-04-5), Granada, Spain, 15 - 19 July 2008

Declaration/Selbständigkeitserklärung

Hiermit erkläre ich, dass diese Arbeit bisher von mir weder an der Mathematisch-Naturwissenschaftlichen Fakultät der Ernst-Moritz-Arndt-Universität Greifswald noch einer anderen wissenschaftlichen Einrichtung zum Zwecke der Promotion eingereicht wurde.

Ferner erkläre ich, dass ich diese Arbeit selbständig verfasst und keine anderen als die darin angegebenen Hilfsmittel benutzt habe.

....